Mathematics Instruction in Dual Language Classrooms

A volume in
Research in Bilingual Education
Mileidis Gort, *Series Editor*

Research in Bilingual Education

Mileidis Gort, *Series Editor*

Innovative Curricular and Pedagogical Designs in Bilingual Teacher Education: Bridging the Distance with School Contexts (2022)
Cristian R. Aquino-Sterling, Mileidis Gort, and Belinda Bustos Flores

Effective Educational Programs, Practices, and Policies for English Learners (2014)
Liliana Minaya-Rowe

The Bilingual School in the United States: A Documentary History (2012)
Paul J. Ramsey

ESL, EFL, and Bilingual Education: Exploring Historical, Sociocultural, Linguistic, and Instructional Foundations (2010)
Lynn W. Zimmerman

Negotiating Social Contexts: Identities of Biracial College Women (2007)
Andra M. Basu

Teacher Training and Effective Pedagogy in the Context of Student Diversity (2002)
Liliana Minaya-Rowe

Mathematics Instruction in Dual Language Classrooms

Theory and Research That Informs Practice

edited by

Marco A. Bravo
Santa Clara University

Kip Téllez
University of California–Santa Cruz

INFORMATION AGE PUBLISHING, INC.
Charlotte, NC • www.infoagepub.com

Library of Congress Cataloging-in-Publication Data

A CIP record for this book is available from the Library of Congress
http://www.loc.gov

ISBN: 979-8-88730-703-9 (Paperback)
979-8-88730-704-6 (Hardcover)
979-8-88730-705-3 (E-Book)

Printed in the United States of America

CONTENTS

FOREWORD

In our joint missions to counteract deficit mindsets and lack of opportunities for emergent bilingual learners (EBLs) in our nation's schools and classrooms, the authors in *Mathematics Instruction in Dual Language Classrooms: Theory and Research that Informs Practice* offer deep insights into the operationalization of epistemological perspectives for our field. Bravo and Téllez have curated and/or co-authored an impressive collection of chapters that are conceptually and pragmatically organized around three central themes:

- "Language, Literacy, and Culture in the Mathematics Classroom: Research-Based Practices"—These four chapters interweave mathematics instruction with disciplinary notions of discourse, policy and pragmatics. The empirical studies in the section are grounded in robust theoretical frameworks and take place in schools and district across the country.
- "Supporting Dual Language Teachers With Mathematics Instruction"—The authors in this section document pre-and-inservice bilingual/dual language teachers' beliefs and practices as a result of their learning experiences. Two of these studies were part of a national professional development program, the results of which are always important to document and disseminate broadly, particularly given the dearth of research in mathematics and dual language education.

Mathematics Instruction in Dual Language Classrooms, pages vii–ix

- Family Engagement in Mathematics Teaching in DLP—Pointing to the critical importance of "co-empowerment" as a result of the reciprocal relationships between educators and parents, the authors of the two chapters in this section highlight how their can studies inform better communication and connections home and school can contribute to childrens' mathematical, linguistic and cultural identities.

The three complementary sections of this book provide dual language programs, their educators, families/communities and emergent bilingual learners with the very much needed research to support and enhance our current understandings of contemporary issues in mathematics instruction in dual language programs.

According to the most recent report of the National Academy of Arts, Sciences and Mathematics (NASEM, 2018), *English Learners in STEM Subjects* (2018), creating systems-and classroom level supports that recognize the assets that EBLs contribute to classroom learning require that educators provide access and equity in designing and implementing program models, curriculum and instruction. However, for EBL students, research has shown that the lack of access to quality mathematics instruction has been linked to decreased student engagement, overemphasis on standardized testing, lack of higher-level course access for secondary level ELs, increased achievement gaps and poor critical thinking abilities—in essence, a lack of equity in access and outcomes (Umansky, 2016). In this case of mathematics instruction in dual language programs, the authors elevate the translinguistic repertoires, cross-and-intercultural and multilingual pedagogic mathematical knowledge through research and practice in a variety of contexts. To better support student access, opportunity and performance, researchers and educators have emphasized the need to develop students' cognitive, academic, multilingual, social, and emotional development (NASEM, 2017), and yet this has yet to be realized in our nation's schools.

This book considers the shifts needed to change the school quality narratives to uplift EL/ML's and their teachers' "linguistic [and mathematical] genius" (Alfaro & Bartolomé, 2018). In *Mathematics Instruction in Dual Language Classrooms: Theory and Research that Informs Practice,* Bravo and Téllez and their colleagues demonstrate ways in which the research, theories and practices in dual language programs amplify the language(s) *of* mathematics and the language(s) *in* mathematical discourses/exchanges and vice versus. They also signal a need to continue to build a research agenda that addresses the mathematical, linguistic and cultural educational ecologies in the best ways that our children deserve.

—Magaly Lavadenz

REFERENCES

Alfaro, C., & Bartolomé, L. I. (2018). Preparing ideologically clear bilingual teachers to recognize linguistic geniuses. In A. Wagner, V. M. Poey, & B. R. Berriz (Eds.), *Art as a way of talking for emergent bilingual youth* (pp. 44–59). Routledge.

National Academies of Sciences, Engineering, and Medicine. (2018). *English Learners in STEM subjects: Transforming classrooms, schools, and lives.* The National Academies Press.

Umansky, I. M. (2016). Leveled and exclusionary tracking: English learners' access to academic content in middle school. *American Educational Research Journal, 53*(6), 1792–1833.

INTRODUCTION

SPEAKING THE LANGUAGE OF MATHEMATICS IN DUAL LANGUAGE SETTINGS

Ninety/Ten or 10/90 or 110010/110010? What we need to know about mathematics instruction in dual language programs.

The delusion is extraordinary by which we thus exalt language above nature: making language the expositor of nature, instead of making nature the expositor of language. (p. 55)

—A. B. Johnson (1836)

A Treatise on Language: Or, the Relation Which Words Bear to Things, in Four Parts

Bilingualism is my superpower.

—Slogan adopted by the National Association or Bilingual Education

We have sub-titled our introductory chapter to this volume with a tongue-in-cheek reference to the ratio of native/target language percentages often considered in dual language programs (DLP). (Because we love mathematics, the third ratio is presented in binary numbers.) But we and the authors in this volume have a very serious purpose. As educators foremost and researchers by training, we often hear DLP advocates talking about their programs in these very general "ratio" terms. But the issues facing DLP

Mathematics Instruction in Dual Language Classrooms, pages xi–xxvii educators in the teaching of mathematics are more complicated. For instance, the most straightforward of questions for DLP educators—"In what language do we teach mathematics in a DLP?"—raises important pedagogical, policy, curricular, and personnel concerns. What advantages might we gain by beginning mathematics instruction in the native language of the students? DLP educators typically begin their students' literacy instruction in the native language, so what about mathematics?

One point we've heard made by some DLP educators is that it doesn't matter: "Mathematics is a 'language-free' discipline," they argue. This view, however, is entirely wrong. As long as there has been serious mathematics, we've known that mathematics cannot happen without language (Bullock, 1994). Other educational theorists have made the connection even clearer (Pimm 1981; Spanos et al., 1988). We have come a long way from viewing mathematics as a "language-free" discipline and are now in a space where we acknowledge the existence of a mathematics "register." For instance, the language of mathematics holds many technical concepts to learn (i.e., vocabulary), in addition to long noun phrases, which are often part of a nominalization. Dual meanings are common for mathematical vocabulary. Complex phrases are often embedded in long mathematical sentences. The use of the passive voice, sentences containing multiple clauses, interrogatives, and many conditional statements require linguistic awareness to decipher intended meanings in mathematics. Finally, complex discursive practices inform mathematical reasoning and argumentation. The reasoning required of a student includes knowledge of discourse patterns, or a logical explanation that is justified with evidence and offers a critique of the approach taken to solve the mathematical problem. It is not hyperbole to say that learning mathematics is tantamount to learning language.

LANGUAGE OF MATHEMATICS

Below, we describe more specific examples of the language of mathematics and offer instructional examples.

Mathematical Vocabulary

Structured vocabulary instruction in mathematics is absolutely necessary in order for DLP students to understand their mathematics textbooks. In fact, the correlation between vocabulary knowledge and comprehension of content area texts is well documented (Blachowicz & Fisher, 2000). Moreover, precision with vocabulary is valued in mathematics and is central to the Common Core State Standards for Mathematics (2010). Students are

TABLE I.1 Everyday/Mathematical Word Chart

Everyday Word	Mathematical Word
Problem	Expression
Find Answer	Calculate
Same	Equivalent
Break down	Decompose

likely to be more familiar with the everyday synonym for the more academic and more precise mathematical term. For example, students may know the term "answer" and be less familiar with the term "product." While these two terms may seem perfectly synonymous, the term product connotes a response from a particular type of mathematical problem, multiplication. But this example is just one among many. See Table I.1 for additional examples.

Students learn precision with mathematical vocabulary often by beginning with words and phrases they already know and then move toward the mathematical term, a process known as word consciousness (Scott & Nagy, 2004). Mathematical vocabulary is also represented in long complex noun phrases (Wong-Fillmore, 2007). Take, for example, the noun phrase "Identity Property of Addition $(0 + 3 = 3;\ 3 + 0 = 3)$," if a student tries to read this phrase word by word, they could lose the meaning being represented by the entire noun phrase. Even a simple noun phrase as "bar graph" can cause confusion if the term graph is read as a verb rather than as part of the noun phrase connected with the term "bar." Instruction on noun phrases can include providing circles around the noun phrase to signal to students to read together.

Mathematical Syntax

In analyzing the linguistic complexity of mathematics items on a state exam, Martinello (2008) noted a particular type of syntax being utilized to explain word problems. Sentence length, number of prepositional phrases, embedded clauses, and use of passive voice was noted in the math problems. Solano-Flores et al. (2013) note in their analysis of the Smarter Balanced assessment items word problems that were in both interrogatives and conditional grammatical form within the same problem. They provide the example, "How many flowers can Jane buy with $7.55 if each flower costs 76 cents?" In such items the linguistic load is heavy and can take cognitive energy from the conceptual understanding that the item is attempting to elicit. Conjunctions, similarly construct new meanings in the syntax of mathematics, linking clauses in very precise ways (Schleppegrell, 2007).

Conjunctions like "when" and "hence" in the following problem functions to link the two statements:

> To find the range let us take certain values for x;
> When $x = 6$: a and b is false. Hence $a \wedge b$ is false.

These terms "when" and "hence" take on precise meaning that if overlooked as one might if these terms were found in a narrative text, understanding the problem and arriving at the appropriate solution becomes more challenging.

Finally, we want to mention the polysemous nature of mathematical vocabulary everyday meanings (e.g., "Operation"—what a doctor does to help sick people) versus the specific mathematical meaning (e.g., "Operation"—a mathematical process like addition and subtraction). Students must understand the differences between these terms, and if they are emerging bilingual learners, or EBL,[1] they will also need to negotiate terms in their native language.

Mathematical Discourse

The speech community of mathematicians is bound by certain rules when explaining or debating mathematics (see Solis, this volume). Explaining and justifying one's own mathematical thinking are critical components of the Common Core State Standards–Mathematics (National Governors Association Center for Best Practices & Council of Chief State School Officers, 2010). A mathematical explanation makes sense of a mathematical concept and illustrates procedural fluency, application of mathematical procedures accurately. The explanation begins with a statement of the concept (a positive number added to another positive number equals a positive number; Mueller & Maher, 2009). The statement is not in dispute nor is their doubt of the concept. It is followed with a set of facts or descriptions that demonstrate a causal relationship, explaining how the mathematical concept came to be understood.

Argumentation in mathematics involves making a claim, building reasoning to support their claim, evaluating the reasoning of others, and a closure that responds to the truth or falsity of the mathematical case (Knudsen et al., 2017). This is truly different from an explanation given the claim is meant to be justified. In Table I.2 we provide examples of how teachers we have worked with have scaffolded students in developing mathematical argumentation.

These constructs, vocabulary, syntax, and discourse, demonstrate clearly the relationship between mathematics and language and the linguistic

TABLE I.2 Argumentation

Problem. A student in your class claims that the sum of any three consecutive counting numbers is divisible by 3.	
Claim I agree I disagree	
Reasoning Use equation State math rule(s) (e.g., theorem, postulate) Draw representation of your math thinking	
Evaluate other's reasoning They are right/wrong because… They could be right/wrong because…	
Closure. Restate claim	

knowledge mathematics requires. It therefore matters greatly how DLP educators approach the teaching of mathematics. We know that all public school DLP teach mathematics, but we know very little about the mathematics achievement of students in these contexts. Existing research has begun to offer insights, but the findings are tentative (e.g., Valentino & Reardon, 2015). In any case, the stakes for our EBL in DLP are high. Data from the National Center on Educational Statistics shows that EBLs generally score about a standard deviation lower on tests of math achievement than their native English-speaking counterparts. Overall, fourth and eighth grade scores on the National Assessment of Educational Progress (NAEP) math assessment decreased in 2022 and EBLs continued to score among the lowest, and from 2013 to 2022 the gap in the scores for native English Speakers and EBLs did not change significantly (NAEP, 2022). The NAEP 2022 results show 29% at or above proficient for native English-speaking students compared to 4% for EBLs in eighth grade mathematics.

The academic underperformance of EBL is more than the consequence of needing to learn English (or whichever the dominant language is in the DLP). Cultural factors are obviously in play. For instance, we cannot ignore the fact that for many EBL families, mathematics and language are used differently in the home (see Stoehr, Bravo, this volume) than what is expected at school. These differences include the use of mathematics during cooking with a focus on measurement or such language practices as writing letters to family members in their country of origin. DLP that bridge this divide between home and school have shown efficacy to support EBL learning outcomes (Civil 2007). Based on the positive results of such models, parent engagement activities now take on a less-deficit orientation toward family approaches to language and mathematics learning (Sarmiento-Quezada, this volume).

At the intersection of language and culture is the value educators place on the native languages of the students participating in a DLP. For centuries, it was not uncommon for teachers and others to discourage or even punish the use of students' native language in school. Often, educators believed that students' native languages were unsophisticated and incapable of expressing complex thoughts and tried to curtail their use. In the early 20th century, Franz Boas (Jakobson & Boas, 1944) and others provided evidence that languages (or cultures) cannot be "ranked" on any scale, and yet incorrect views of this sort persist.

Because we still find some educators discouraging the use of students' native languages, or at least undervaluing them, a modern version of Boas' project has emerged in a concept known as "translanguaging." Relying on theories suggesting that bilinguals do not possess two separate systems for languages but rather access a single "well" or linguistic repertoire to make meaning (cf. Grosjean, 2012) and a renewed interest in raising the status and value of students' native language, in their minds and those of others (Araújo Dawson & Williams, 2008). As noted, the translanguaging "movement" (García, 2009; García & Wei, 2014) can be traced to previous efforts promoting the value of all languages and the removal of any shame that may accompany the use of a "low-status" language, as well as encouraging the mixing of languages to allow speakers to display their knowledge of the world as fully as possible.

For our part, we are interested if translanguaging can help learners to make sense of subject areas like mathematics (see Castillo, this volume). When permitted to translanguage, evidence suggests students engage in tasks more readily and are also able to leverage their full linguistic repertoire to acquire teacher's learning goals (Pontier & Gort, 2016; García et al., 2017). There is strong evidence that this can support students' vocabulary development (Pontier & Gort, 2016), academic presentations (Garcia et al., 2011), reading comprehension (Garcia et. al., 2017), and deeper content understanding (de Jong & Bearse, 2014; Rodríguez-Valls et al., 2017).

In our own work, we have promoted a translanguaging strategy that assists teachers of EBLs to devote instructional attention to cognates during mathematics instruction (Bravo et al., 2022). In our work with bilingual preservice teachers, we share rules for identifying cognates across Spanish and English, including the rule that if a Spanish word ends in *-ción* it is likely an English cognate that ends in -tion (Montelongo & Hernandez, 2023). Examples include *división*/division, *ecuación*/equation. At times the Spanish word or English word may be more salient for the student, depending on the frequency of engagement with the particular word, yet for EBLs, they can access the meaning of the word either through their native language or

English. Moreover, in our work of supporting teachers of EBLs, we noted a certain type of cognate that occurs in mathematics. Some English/Spanish cognates are everyday words in Spanish (e.g., *igual, dividido, volumen*) but more academic terms in English (equal, divide, volume). In these instances, EBLs linguistic repertoire provides them with unique access to academic meanings using either their English or Spanish language. While a very useful strategy, it is a strategy that requires instruction. Garcia et al. (1998) noted that EBLs did not automatically use this strategy to make sense of unfamiliar English words, but did so only after explicit and systematic instruction was provided. We also recognize that cognates are rare between English and many other non-Latin-based languages (e.g., Mandarin, Vietnamese), so teachers working with speakers of these languages will not be able to rely on cognates.

Teachers of EBLs who promote translanguaging practices tend to create a linguistic environment that honors and leverages students' full linguistic repertoire (Celic & Seltzer, 2013; García et al., 2017; Mazak & Carroll, 2016). These practices break from previous instructional traditions where it was taboo for teachers and students to use both languages (see Mendoza, this volume). The model was to keep the languages separated with the goal of not confusing students (Ramirez et al., 1991). It was feared that students would become "linguistically lazy" and not try to develop fluency in the target language. Current research on translanguaging illustrates that students become "linguistically vibrant" as they leverage their understandings of two codes to negotiate and make meaning (Creese & Blackledge, 2015; DeNicolo, 2019; Durán & Palmer, 2014).

We began this chapter discussing the relationship between mathematics and language but have found ourselves in a discussion about cultural values, language preferences and respectfulness. This should come as no surprise, but why are decisions about language cultural at their core? The following section explores this question while offering connections to DLP.

CULTURE AND LANGUAGE

Would the languages of people exist if there were no people? This is a preposterous question, but it forces us to consider language as a product of culture, and not the other way around. Some of us might agree with this view, but only to a point. Doesn't our language shape the culture or even our overarching thinking? After all, haven't we all heard the Inuit "words for snow" story? The basic myth is that Inuit speakers have many more words for snow than English speakers, thus offering evidence that our culture and even our "reality" is shaped by language. Ever since the myth became popular, linguists and anthropologists alike (e.g., Pullum 1991) made efforts to discredit it. They have argued that Inuit speakers needed more words for snow because their environment required finer distinctions. Language does not create variations in thinking or culture. Cultural forces make language do its work.

Other explanations about the persistence of the myth suggest a confusion regarding what counts as a word. If speakers of many different languages were confronted with the need to describe different kinds of snow, the task is simple: use adjectives. In English, we might say, "Hey, that's wet snow. We might have trouble moving it." And if we needed to describe wet snow over and over, the terms might merge into a new word: "wetsnow." Speakers of some languages permit this new word creation more quickly than others, but every language (i.e., every speaker) can do it. Languages vary, but their variation is a consequence of what the culture requires of them. Languages are not people; they don't do anything.

If we can agree with the premise that languages are the product of culture, we are off to a good start. But more confusion may follow: We often conflate the social function of language and languages as they are represented in a social world, a world in which people place a value on languages based upon the people who speak that language or dialect. For example, we can turn to the myth that French is the most romantic of all human languages. For many English speakers, if someone whispers "*Je t'aime*" to you, it must mean that they really, really love you. However, French is no better at expressing love than any other language, but French culture is regarded as romantic, so the language gets confused with the people who speak it. Those who speak German or Hindi or Tagalog are just as capable of deep, loving emotions and can use their language to say so.

So, what does this have to do with mathematics teaching and learning in DLP? First, any instructional decisions regarding language are a function of culture. In fact, the decision to teach mathematics at all is rooted in culture and history. Of course, it would be difficult to imagine that at any time in the near future we might decide to stop teaching mathematics in U.S. schools, but there is nothing preventing it. However, we do not need to resort to such extreme hypotheticals to demonstrate the cultural foundations of mathematics education. Rather, comparative data aids in understanding. For instance, we might use the number of minutes per week mathematics is taught in various countries as a measure of the value placed on mathematics. Admittedly, countries are not cultures, but the cultures that comprise a country are bound by important formal rules and codes and thus connected in many ways. International data suggests that countries such as Singapore and England spend about 10% more instructional time devoted to mathematics than the world average. The number of minutes mathematics is taught in the United States is the same as the world average (Elliott et al., 2019; Leung, 2014). These studies also show that students place varying degrees of importance on learning mathematics and holding a positive attitude towards mathematics. Students in Egypt and Iran lead other nations in these two constructs; Singapore and the United States are not far behind. Interestingly, students in Japan and Korea rate near the bottom in positive attitudes towards mathematics. Measures of adult attitudes, not surprisingly, reflect the student data. But no country seems to think mathematics is unimportant; they all teach it, and almost always teach it in the language that serves as the medium of instruction for other subjects. Again, because mathematics, like any other subject, can be taught in any language, we find a cultural choice.

Even in those instances where we believe that language is the reason behind differences among people, we find that culture to be the root cause. One line of intriguing research into the differences in mathematics achievement addresses the role of number-naming systems across languages, Chinese and English in particular. Researchers studying number naming often begin by noting international data showing that students in Asian countries score higher on international tests of mathematics than do children in English speaking countries (Mullis et al., 2004). These data are used by a wide variety of policy-makers and others, sometimes to show that the United States, while spending more money per student on schools than many Asian countries, does not get a return on that investment. But forget about the anti-public school and often anti-teacher union tenor of the discussion and ponder the genuine reason for the disparity. Do we assume that children in Asian countries are simply smarter than students in the United States? Do we blame the disparity on the tests themselves, as "biased" against students in the United States?

Both of these specious claims fall apart on closer examination, so why not consider the difference in achievement as a feature of culture? This is the goal of number naming research. Briefly explained, Asian languages use number naming systems that conform to a clear base 10 understanding, unlike English that uses "eleven" and "twelve." Chinese follows a consistent base-10 rule. In fact, a literal translation of the Chinese name for 11 is "ten one." In addition, Chinese number names from 100 to 109 (and 200 to 209, etc.) interpose a term, roughly translated to 0, to represent the absent value (e.g., 105 is said "one hundred, zero, five"; 10,005 is said "ten thousand, zero, zero, five"). English, by contrast, uses the word "and" (e.g., "one hundred *and* five" or "ten thousand and five"), thereby ignoring place value. For decades, studies have shown that Chinese speaking children learn to count more quickly than English speaking children (Miller et al., 1995); therefore, it is possible that the difference in mathematics achievement by country be owing in part to how English mixes base 12 and base 10 systems and obscures place value in numbers beyond 100.

We might be compelled to consider this difference entirely the result of differences in languages, but language differences always arise from cultural differences, and the case of number naming is no different. The full details of the story require more space than we have to share, but the records indicate that early Germanic languages, on which English is based, did not have numbers above 10. Eleven and twelve were used to refer to "many" and "more than many" (Menninger, 2013). And with early Christian texts using these numbers often, they became common terms to refer to more than 10, only later to be recognized as the numbers we use as 11 and 12. When early Anglo-Saxon English speakers were forced to adopt Arabic numbers (likely for trading purposes), the language and number naming system changed, adding the "teens" and base 10 number naming beginning at 20.

Other languages such as Spanish are similar to English ("once, doce, trece, catorce") and conform to base 10 counting at 16 (dieciséis). Interestingly, the Mayan number system uses base 20 throughout, which we might speculate would produce precocious mathematicians in our modern era, but because Spanish "explorers" decimated the Mayan people and their language in the 1500s, we'll never know. In fact, most number systems have been merged or pieced together, while others were simply destroyed to suit the preferences of those in power. Our overall point is, as Chrisomalis (2010) notes: "The role of various *social factors* (emphasis added) in explaining the history and development of numerical notation systems differs from case to case, depending on historical context, but they are always there (p. 401).[2]

Returning to the differences in mathematics scores, English speaking students of course eventually catch on to base 10, but the initial confusion may result in lasting underperformance when compared to Chinese speakers. Additional research might yield a definitive answer, but any explanations for the score differences are cultural. For example, some have argued that differences in achievement between Chinese and other Asian students and U.S. students in mathematics is the result of the value placed on learning in Confucian cultures (Tan, 2013). Or perhaps it is the importance of testing in the history of China (Song, 2016). But like the number naming systems, these are *cultural* differences, not of intelligence, nor of language.

Returning to one of our original questions, the choice of languages to teach in schools is an obvious example of a cultural preference: Any subject can be taught in any language if teachers know the language. And in most cases, the medium of instruction (MoI) in schools is the "official" language of the people in power. And in the United States, except for rare historical examples (see Kloss, 1977/1998), the MoI has been English and only English. Of course, secondary schools in the United States typically require some courses in learning a language other than English (LOTE) for graduation, and most universities require 2–4 years of study in a LOTE for admission. Surprisingly, these LOTE requirements have been in place for many decades and have faced surprisingly little controversy. So to argue that the United States does not value the speaking of a LOTE is not strictly accurate, but the moment one proposes that students learn school *content* in a LOTE (i.e., the MoI is not English) that's when trouble starts, especially when the learners are young and yet to gain fluency in English. This is the heart of the controversy around programs such as bilingual education, and the source of ham-handed reforms such as Proposition 227 in California (Garcia & Curry-Rodríguez, 2000). This history of MoI efforts in the United States is important to understand, but we do not have the space to review the long history of language education in the United States and instead refer readers to excellent examples (e.g., Gándara & Escamilla, 2016; Ovando, 2003; San Miguel, 2004).[3]

What do we know now about the MoI in mathematics teaching in DLP? Not much. Unless the language of instruction is mandated in DLP, as it is, for example, in the state of Utah (Utah Dual Language Immersion, 2018), it appears to vary. In an investigation conducted by one of our colleagues, Balloffet (2020) surveyed 41 California principals working in DLP schools and found a wide range of programming models and rationales for these choices. The principals described 13 different programming models for which language was used for instruction at each grade level. These were sorted into four categories: schools in which the MoI is English ($n = 4$), schools in which the MoI is Spanish ($n = 8$), those teaching mathematics in both languages ($n = 10$), and schools that switched LoI from Spanish to English in the upper grades ($n = 19$), generally around the 3rd grade. The survey found that some schools decided the LOTE mathematics curriculum was good at the lower grades but weakened as the students progressed. Other schools reported that teachers lacked confidence teaching mathematics at the upper grades. But whatever the decisions, the decisions were made based on cultural (e.g., economic, personnel) concerns.

The clear goal of any DLP is to educate "balanced bilingual" speakers. Knowing two languages can result in metalinguistic advantages or, as many DLP parents suggest, increased employment opportunities and perhaps a better salary later on in life, although researchers have largely refuted this claim (Fry & Lowell, 2003). But the real reason for knowing more than one language is social through and through. Our history as a species suggests that the only reason to learn another language was to be able to interact with others who speak that language. And the only reason to learn to do mathematics in a specific language is to "do mathematics with people who also do mathematics in that language." But it's easy to be convinced of other reasons. For instance, returning to our number naming example, we might be inclined to teach children mathematics in any of the Asian languages that promote base 10 and a deeper understanding of place value. But remember that there are likely other reasons why Asian students outperform

U.S. students. Teaching mathematics in Chinese will not automatically result in improved mathematical performance.

If we decide to teach mathematics in Mandarin in a DLP, the purpose must be because we want the students to be able to do mathematics with others who do mathematics *in Mandarin*, with the goal of creating cultural understanding. Building on our earlier example, imagine a fourth grade DLP classroom learning place value in Mandarin using its number naming system. The teacher has invited one of the students' grandmothers to help in small groups. Whatever mathematics the students learn from the grandmother is secondary to what they learn about her as a person, a cultural being. They might notice that her Mandarin dialect is a little different from their teachers' and wonder why. She might share how mathematics was taught when she was a young girl growing up in China. They might notice the clothes she wore or how kind she was. The non-native Mandarin speaking students might have seen elderly Asian people in their town or portrayed in the media, but now they *know* one. Yes, they are learning place value, an important goal in mathematics education, but doing mathematics with someone you did not know beforehand is more important. To put it plainly, if educators working in a DLP have goals beyond cultural growth and understanding, they are likely to be disappointed (Martin-Beltrán, 2010; Téllez, 2010).

CONCLUDING THOUGHTS

As the NABE slogan quoted at the start of our chapter suggests, bilingualism is a "superpower" DLP educators can confer upon their learners. EBLs' bilingualism can serve as a superpower to aid them in understanding the language of mathematics, whether they confront mathematics in their native language or English. Utilizing their superpower will not only enhance their mathematical learning, but also give them an opportunity to sharpen their bilingualism proficiency, enhancing their college and career prospects.

In a final exploration of the relation between language and culture, we turn to Chomsky (1990), who imagines a world where speakers of a particular language consistently misuse a term, then all dictionaries vanish. In this new world, the misuse of the term would become accepted and newly created dictionaries would repeat incorrect usage. Chomsky uses this fictional example to argue that in order to understand the "science" of language, we must clarify the "concept of 'community norms' or 'conventions'" (p. 509), which he argues are not at all clear. This is the contested, even unexplored, space in which DLP do their work. What role do languages play in creating communities or solidarity among speakers and among content areas like mathematics? For DLP educators, who seek to create a linguistic community that is accepting of students' linguistic identities and supports EBLs' sense of belonging (de Jong et al., 2020), the challenge is great. DLPs are positioned to take advantage of the possibilities offered by the synergy between language and mathematics, both superpowers, in fact, and create a linguistic ethos that supports all learners.

PURPOSE, LIMITATIONS, AND ORGANIZATION OF THE BOOK

We decided to create this book because we believed that DLP educators, including ourselves, had not considered how mathematics and language could work together in a DLP to enhance learning in complex cultural contexts. The chapters we have assembled provide provisional answers to some of the important questions, but after reading them all we admit we are left with more questions that we started with, which is not necessarily an unwanted outcome. And we freely admit that the book may disappoint readers seeking prescriptive guidance on how to teach mathematics in DLP. Although there are plenty of pedagogical and programmatic suggestions to consider, all require a filtering through of each DLP' specific context (e.g., availability of curricular materials in the language of instruction).

BOOK ORGANIZATION

This volume is organized around three themes. The first section, "Language, Literacy, & Culture in the Mathematics Classroom: Research-Based Practices," addresses questions regarding the learning of mathematics in DLP classrooms. The chapters explore pedagogical considerations when teaching mathematics to EBLs. For example, the chapter authored by Castillo investigates translanguaging in mathematics and the results relate directly to our argument regarding the role of culture in DLP.

The second section, "Supporting Dual Language Teachers With Mathematics Instruction," focuses on teacher beliefs and practices regarding the integration of mathematics and language. The chapter authored by Pathoff for example, explores an innovative approach to teacher professional development and one uniquely suited to assist DLP teachers to learn more about integrating language and mathematics, but the model might also be effective in other highly specialized teaching contexts.

The third section, "Parent & Community Engagement in Mathematics Teaching in DLP" provides insights into models of engaging parents but more importantly on how to elicit the mathematical knowledge and language skills parents possess to leverage this knowledge to enhance student learning. The two chapters in this section, Stoehr and Bravo and Sarmiento-Quezada, emerge from efforts to support new teacher learning regarding parent engagement.

SPECIAL NOTE

Readers of this volume will notice that several of the chapters reference a funded project known as Mathematics, Language, and Literacy Integration in Dual Language Settings (MALLI). MALLI was a National Professional Development grant funded by the Office of English Language Acquisition (OELA) in the U.S. Department of Education. We were the primary principal investigators (i.e., directors) of the project. (See https://malli.sites.ucsc.edu/ or https://www.scu.edu/ecp/centers-and-partnerships/ for more information.) The primary mission of MALLI was to assist DLP teachers to integrate language processes (literacy, discourse, and vocabulary) into their mathematics instruction, regardless of the language of instruction. MALLI also assisted preservice and first-year teachers to learn more about mathematics and language integration, as well as helping parents to better assist their students in mathematics. Many of the chapter authors worked on the MALLI project at some point during its 6 years.

We would also like to thank Mileidis Gort, the series editor, who helped us sharpen the focus and provided superb support throughout. We are also grateful to the copy editors at Information Age who made each and every chapter better. And finally to our index editor, whose quick and exacting work "de-stressed" us.

—Kip Téllez
Marco A. Bravo

NOTES

1. We have adopted the term emerging bilingual learners (EBL) to represent many of the learners in DLPs, although we understand this term is not strictly accurate (some students might be learning additional languages) nor widely accepted (some educators prefer multilingual learners). For a wider discussion, see Dektor and Téllez (2023).
2. See https://www.utahdli.org/instructional-model/ for an overview.
3. Menninger makes a similar point: "A people's number sequence is not a system created fresh out of the pure workings of the mind; it is rooted in the same soil as the people. Like culture itself, it grows up slowly over the millennia, and even in its mature form it reveals the history of its people through the successive deposits of the passing years" (p. 162, e edition).

4. We anticipate that many readers of this volume will already be familiar with this history.

REFERENCES

Araújo Dawson, B., & Williams, S. A. (2008). The impact of language status as an acculturative stressor on internalizing and externalizing behaviors among Latino/a children: A longitudinal analysis from school entry through third grade. *Journal of Youth and Adolescence, 37*, 399–411.

Balloffett, L. (2020). *Mathematics instruction in California's dual language programs: Language of instruction and principal decision-making* [Unpublished manuscript].

Balloffet, L., & Téllez, K. (2021). How are California's Latina/x/o students faring?: Charter elementary schools' Spanish/English dual language programs. *Journal of Leadership, Equity, and Research, 7*(2), n2.

Blachowicz, C. L. Z., & Fisher, P. J. L. (2000). Vocabulary instruction. In M. L. Kamil, P. B. Mosenthal, P. D. Pearson, & R. Barr (Eds.), *Handbook of reading research* (Vol. 3, pp. 503–523). Erlbaum.

Bravo, M. A., Mosqueda, E., & Solís, J. L. (2022). Preparing teachers for dual language contexts: Strategies for teaching and assessing mathematics in two languages. In M. Machado-Casas, S. I. Maldonado, & B Bustos Flores (Eds.). *Evaluating bilingual education programs: Assessing students' bilingualism, biliteracy and sociocultural competence.* Peter Lang International Academic Publishers.

Bullock, J. O. (1994). Literacy in the language of mathematics. *The American Mathematical Monthly, 101*(8), 735–743.

Celic, C., & Seltzer, K. (2012). *Translanguaging: A CUNY-NYSIEB guide for educators.* The Graduate Center at The City University of New York. https://www.cuny-nysieb.org/wp-content/uploads/2016/04/Translanguaging-Guide-March-2013.pdf

Chomsky, N. (1990). Language and problems of knowledge. In A. P. Martinich (Ed.), *The philosophy of language* (2nd ed.; pp. 509-527. Oxford University Press.

Chrisomalis, S. (2010). *Numerical notation: A comparative history.* Cambridge University Press.

Creese, A., & Blackledge, A. (2010). Translanguaging in the bilingual classroom: A pedagogy for learning and teaching? *The Modern Language Journal, 94*(1), 103–115.

de Jong, E. J., & Bearse, C. I. (2014). Dual language programs as a strand within a secondary school: Dilemmas of school organization and the TWI mission. *International Journal of Bilingual Education and Bilingualism, 17*(1), 15–31.

Dektor, R., & Téllez, K. (2023). Who are the students learning English and how do we talk about them? In A. Esmail, A. Pitre, A. Duhon-Ross, J. Blakely, & B. Hamann (Eds.), *Social justice perspectives on English language learners* (pp. 123–135). Hamilton Press.

DeNicolo, C. P. (2019). The role of translanguaging in establishing school belonging for emergent multilinguals. *International Journal of Inclusive Education, 23*(9), 967–984.

Durán, L., & Palmer, D. (2014). Pluralist discourses of bilingualism and translanguaging talk in classrooms. *Journal of Early Childhood Literacy, 14*(3), 367–388.

Elliott, J., Stankov, L., Lee, J., & Beckmann, J. F. (2019). What did PISA and TIMSS ever do for us? The potential of large scale datasets for understanding and improving educational practice. *Comparative Education, 55*(1), 133–155.

Fry, R., & Lowell, B. L. (2003). The value of bilingualism in the U.S. labor market. *ILR Review, 57*(1), 128–140.

Gándara, P., & Escamilla, K. (2016). Bilingual education in the United States. In O. García, O. Garcia, & A. Lin (Eds.), *Encyclopedia of language and education 5* (pp. 1–14). Springer.

Garcia, E. E., & Curry-Rodríguez, J. E. (2000). The education of limited English proficient students in California schools: An assessment of the influence of Proposition 227 in selected districts and schools. *Bilingual Research Journal, 24*(1–2), 15–35.

García, G. E. (1998). Mexican-American bilingual students' metacognitive reading strategies: What's transferred, unique, problematic? *National Reading Conference Yearbook, 47*, 253–263.

García, O. (2009). *Bilingual education in the 21st century: A global perspective.* Malden/Oxford: Wiley/Blackwell.

García, O., Johnson, S. I., & Seltzer, K. (2017). *The translanguaging classroom: Leveraging student bilingualism for learning.* Caslon.

García, O., & Li, W. (2014). *Translanguaging: Language, bilingualism and education.* Palgrave Macmillan.

Grosjean, F. (2012). Bilingualism: A short introduction. In F. Grosjean & P. Li (Eds.), *The psycholinguistics of bilingualism* (pp. 5–25). John Wiley & Sons.

Jakobson, R., & Boas, F. (1944). Franz Boas' approach to language. *International Journal of American Linguistics, 10*(4), 188–195.

Kloss, H. (1977/1998). *The American bilingual tradition.* Delta Systems and Center for Applied Linguistics.

Knudsen, J., Stevens, H. S., Lara-Meloy, T., Kim H. J., & Shechtman, N. (2018). *Mathematical argumentation in middle school: The what, why, and how.* Corwin Mathematics.

Leung, F. K. (2014). What can and should we learn from international studies of mathematics achievement? *Mathematics Education Research Journal, 26*, 579–605.

Martin-Beltrán, M. (2010). The two-way language bridge: Co-constructing bilingual language learning opportunities. *The Modern Language Journal, 94(*2), 254–277.

Martiniello, M. (2008). Language and the performance of English language learners in math word problems. *Harvard Educational Review, 78*(2), 333–368.

Mazak, C. M., & Carroll, K. S. (Eds.). (2016). *Translanguaging in higher education: Beyond monolingual ideologies.* Multilingual Matters.

Menninger, K. (2013). *Number words and number symbols: A cultural history of numbers.* Dover Publications.

Miller, K. F., Smith, C. M., Zhu, J., & Zhang, H. (1995). Preschool origins of cross-national differences in mathematical competence: The role of number-naming systems. *Psychological Science, 6*(1), 56–60.

Montelongo, J. A., & Hernández, A. (2023). *Teaching cognates/cognados through picture books: Resources for fostering Spanish-English vocabulary connections.* Brooks Publishing.

Mueller, M. F., & Maher, C. A. (2009). Convincing and justifying through reasoning. *Mathematics Teaching in the Middle School, 15*(2), 108–116.

Mullis, I. V., Martin, M. O., Gonzalez, E. J., & Chrostowski, S. J. (2004). *TIMSS 2003 international mathematics report: Findings from IEA's trends in international mathematics and science study at the fourth and eighth grades.* International Association for the Evaluation of Educational Achievement. Herengracht 487, Amsterdam, 1017 BT, The Netherlands.

National Governors Association Center for Best Practices & Council of Chief State School Officers. (2010). *Common Core State Standards for mathematics.* Retrieved from http://www.corestandards.org/Math/Practice

Ovando, C. J. (2003). Bilingual education in the United States: Historical development and current issues. *Bilingual research journal, 27*(1), 1–24.

Pimm, D. (1989). *Speaking mathematically: Communication in mathematics classroom.* Routledge.

Pontier, R., & Gort, M. (2016). Coordinated translanguaging pedagogy as distributed cognition: A case study of two dual language bilingual education preschool coteachers' languaging practices during shared book readings. *International Multilingual Research Journal, 10,* 89–106.

Pullum, G. K. (1991). *The great Eskimo vocabulary hoax and other irreverent essays on the study of language.* University of Chicago Press.

Ramírez, D. J., Yuen, S. D., Ramey, D. R., & Pasta, D. J. (1991). *Final report: Longitudinal study of structured-English immersion strategy, early-exit and late-exit transitional bilingual education programs for language-minority children.* National Clearinghouse for English Language Acquisition. https://ncela.ed.gov/files/rcd/BE017748/Longitudinal_Study_Executive_Summary.pdf

San Miguel, G. (2004). *Contested policy: The rise and fall of federal bilingual education in the United States, 1960–2001* (Vol. 1). University of North Texas Press.

Schleppegrell, M. J. (2007). The linguistic challenges of mathematics teaching and learning: A research review. *Reading & Writing Quarterly, 23,* 139–159.

Scott, J. A., & Nagy, W. E. (2004). Developing word consciousness. In J. F. Baumann & E. J. Kame'enui (Eds.), *Vocabulary instruction: Research to practice* (pp. 201–217). Guilford.

Solano-Flores, G., Barnett-Clarke, C., & Kachchaf, R. R. (2013). Semiotic structure and meaning making: The performance of English language learners on mathematics tests. *Educational Assessment, 18*(3), 147–161.

Spanos, G., Rhodes, N., Dale, T. C., & Crandall, J. (1988). Linguistic features of mathematical problem solving: Insights and applications. In R. R. Cocking, R. T. Cocking, & J. P. Mestre (Eds.), *Linguistic and cultural influences on learning mathematics* (pp. 221–240. Psychology Press.

Song, X. (2016). Fairness in educational assessment in China: Historical practices and contemporary challenges. *Assessment in education: Implications for leadership,* 67–89.

Tan, C. (2013). For group, for self: Communitarianism, Confucianism and values education in Singapore. *The Curriculum Journal, 24*(4), 478–493.

U.S. Department of Education. (2022). Institute of Education Sciences, National Center for Education Statistics, National Assessment of Educational Progress (NAEP), 2022 Mathematics Assessment.

Utah Dual Language Immersion. (2018, January 22). *Utah dual language immersion instructional time.* https://www.utahdli.org/instructionalmodel.html

Valentino, R. A., & Reardon, S. F. (2015). Effectiveness of four instructional programs designed to serve English learners. *Educational Evaluation and Policy Analysis, 37*(4), 612–637.

Wong-Fillmore, L. (2007). English learners and mathematics learning: Language issues to consider. In A. H.Schoenfeld (Ed.), *Assessing mathematical proficiency* (pp. 333–344). Cambridge University Press.

PART I

LANGUAGE, LITERACY, AND CULTURE IN MATHEMATICS CLASSROOM: RESEARCH-BASED PRACTICES

CHAPTER 1

"¿QUÉ VA A PASAR?"

Exploring Middle School Bilingual Students' Algebraic Thinking and Translanguaging in Programming Tasks

Sylvia Celedón-Pattichis
University of Texas at Austin

Carlos López Leiva
University of New Mexico

Phuong Tran
University of New Mexico

Marios S. Pattichis
University of New Mexico

ABSTRACT

This book chapter illustrates how the use of languages (Spanish and English) mediate middle school Latinx bilingual students' and (co)facilitators' explorations of algebraic thinking through computer programming. The participants

Mathematics Instruction in Dual Language Classrooms, pages 3–24

include two groups of 3 students (one predominantly using English and one both languages), a facilitator—typically an undergraduate engineering student—and two co-facilitators who had participated as students learning the first level of the curriculum through an afterschool program. The setting for the afterschool program was in a bilingual middle school that enrolled primarily Latinx students. Data sources include twelve video recordings that lasted 1.5 hours of students' interactions in each group while coding and artifacts such as students' work and curriculum tasks. We draw from one of the twelve sessions in each group that focused on algebraic thinking. Drawing from sociocultural theories of learning, we show how bilingual students manipulated variables through computer programming to make meaning of algebraic concepts that involved exploring variables, evaluating expressions, and generating conjectures. The chapter ends with a discussion and implications for the classroom.

> *We were starting session 2 today, and right now there are two more new students to the group and, well, they all speak Spanish, so it was easier for me since the other two kids cause last week since only one understood Spanish and the other preferred English, it was difficult. But like it's not that hard for me to handle the three of them since like you know they all speak Spanish and they like struggled a lot, but they got it right away. So, it might actually work out really well with them cause they are smart learners, and they could possibly understand the code better.*
>
> —Juanita's Reflection, May 2019

The quote above illustrates how Juanita leveraged her bilingualism in Spanish and English as she co-facilitated the teaching of a curriculum that integrated mathematics and computer programming with an undergraduate engineering student in an after-school program. She refers to the students struggling a lot but, at the same time, "they got it (the content) right away." Her views of bilingual students as "smart learners" and that they can "possibly understand the code better" reflect the central aims of this after-school program.

As the State of Computer Education reports (Code.org Advocacy Coalition & CSTA, 2018), rarely do students enrolled in Title I schools have the opportunity to learn computer programming. Zong (2022) reports that the Latinx population is projected to grow to 111.2 million by 2060, making it 28% of the U.S. population. At the same time, there are approximately 4.5 million multilingual student learners in the United States, with three fourths of this student population who speak Spanish as their first language. This steady growth of the Latinx population and the limited opportunities that Latinx students have in engaging with computer programming has consequences for students like Juanita who attend rural bilingual middle schools with large numbers of English Learners. The majority of students in rural schools tend to be on free and reduced-price lunch and also remain underrepresented in science, technology, engineering, and mathematics

(STEM) fields. Thus, this study sought to address the following research question: "How does the use of language (Spanish/English/translanguaging) mediate middle school Latinx bilingual students' and (co)facilitators' explorations of algebraic thinking through computer programming?"

THEORETICAL PERSPECTIVES

Sociocultural Theories of Learning

Language and culture are instrumental to mediating learning through social interactions (Vygotsky, 1978). Vygotsky foregrounded the importance of play, collaborative learning, guided learning, among other social components, to support how individuals internalize meaning of abstract concepts through the use of mediation tools. Vygotsky (1986) also elaborated on the critical role conversations play in that greater learning occurs where there is collaboration than when an individual works alone. Furthermore, Vygotsky (1986) distinguished between reactive and spontaneous learning in formal and informal contexts, respectively. Reactive learning refers to what most learners experience in formal science education where they react to questions or instructions from instructors and the learning tends to be individually focused. On the other hand, spontaneous learning in informal contexts is focused on offering choices for students to engage voluntarily via their interests and is highly social. According to Murphy (2022), students spend 80% of their time outside of schools where much of the learning is driven by students' interests and curiosity. This learning happens at home, museums, online, after-school clubs, and through digital media and gaming.

Collectively, we draw from sociocultural perspectives of learning to describe how students use language in an after-school program in bilingual middle schools. We also draw from sociocultural perspectives of learning to understand the collaborative work and interactions that happened between bilingual students and (co)facilitators in an after-school program to make meaning of an integrated mathematics and computer programming curriculum.

Languaging, Learning, and Computing in Third Spaces

Sociocultural theories of learning emphasize the importance of "third spaces" for all learners and for bilingual students in particular. "Hybridity and diversity are the building blocks of third spaces," in which conflict and competing discourses often forge new learnings (Gutiérrez et al., 1999, p. 287). In our study, we explore how a third space was created in an afterschool program designed to teach algebraic thinking and computer programming to bilingual (Spanish/English) students, assisted by a native Vietnamese speaking facilitator.

However, given the complex language usage in the program, the third space in our study was further enriched by "translanguaging" (García & Wei, 2014). Translanguaging theorists suggest that bilingual speakers willfully integrate their languages into a coherent meaning making system, whether or not their system follows the rules of either language. The program also encouraged the merging of home and school knowledge, a shared feature of all third space learning. Into this complex learning context, learners in the program used Python, a text-based programming language, to explore new mathematical and computational concepts. While not strictly known as a dual language program (DLP), all participants in this program were encouraged to use their translanguage to explore, explain, and justify the programming choices. Learning mathematics in this linguistic environment mirrors the type of learning we find in DLP examples throughout this volume.

Predicting, Problem Solving, and Learning

In our work, we understand predicting to be tightly connected to spontaneous learning as predicting creates a sense of curiosity just as spontaneous learning can create spaces for imagination, creativity, innovation, and intuitive thinking (Vygotsky, 1986). As a result, prediction increases student participation and supports students' progress from "passive listeners to active thinkers and expand(s) and deepen(s) their mathematical knowledge" (Lim et al., 2010, p. 606). In fact, predicting reflects some of the tensions that can occur in third spaces to expand learning (Engstrom, as cited in Gutiérrez et al., 1999). From this perspective, prediction tasks "can be designed for students to experience cognitive conflicts, resolving which can lead to learning of targeted concepts deeply" (Lim et al., 2010, p. 597). Like Lim et al. (2010), we conceptualize making predictions as playing an important role in constructing knowledge and solving tasks. Prediction involves foretelling or making a declaration before an event happens and allows a space for students to demonstrate their current understanding of mathematical concepts without feeling pressure to commit to certainty in their responses. As such, prediction tasks help uncover students' conceptions in certain topics and they help students "activate and refine their existing knowledge" and to think relationally (Lim et al., 2010, p. 598). Finally, to better support these processes, teachers should use problems that are intrinsically relevant to students. Students are more likely to be engaged in mathematical thinking if they understand and are intrigued by the problem they are to solve.

LITERATURE REVIEW

Mathematics and Computing in Multilingual Contexts

Recent research has shown the importance of integrating computing with other content areas such as mathematics in multilingual contexts (Celedón-Pattichis et al., 2022; Jacob et al., 2022; LópezLeiva et al., 2022; Pozos et al., 2022; Radke et al., 2022; Vogel et al., 2020). These studies have prioritized leveraging holistic dimensions of students' backgrounds, including their languages and cultures. Most importantly, these research studies point to the need to have transrelational methodology for teaching mathematics and computing and center relationships and social construction of knowledge.

Scholars have connected the work on mathematics and computing with translanguaging using sociocultural approaches to teaching and learning mathematics and computing (LópezLeiva et al., 2022; Vogel et al., 2020). For example, Vogel et al. (2020) conducted a study across three bilingual middle schools with the goal of working with educators to implement translanguaging teaching practices to enhance student understanding of computer programming. Findings from this study indicate that affording students opportunities to use their full linguistic repertoires facilitates students' vocabulary expansion and computer programming as a language. Furthermore, students drew from familiar backgrounds of using *telenovelas* (soap operas) to create their own scenarios using Scratch, a programming platform. This translanguaging pedagogical approach using Spanish and English bolstered student acquisition of English vocabulary related to computer programming and coding (Vogel et al., 2020).

Through the same aforementioned research project, Radke et al. (2022) collaborated with teachers to co-design and implement a culturally relevant lesson on post-Hurricane María from Puerto Rico so that students developed computational models of migration. It is important to note that teachers and researchers created an environment where "syncretic literacies" and "translanguaging practices" were used to advance student's understanding of computer programming while challenging dominant monoglossic language learning ideologies. Similarly, the findings point to the need to provide learning spaces where translanguaging can be used to construct, make meaning, justify, explain, and interpret computational models of migration. Implications of this study call for intentional use of translanguaging and syncretic literacies for meaning-making to provide new possibilities in STEM education that may include but are not constrained by disciplinary content. Collectively, these studies share commonalities in that teachers and researchers challenged the status quo by addressing inequities for students who remain at the periphery of STEM education.

INTEGRATING COMPUTING THROUGH ALGEBRA

Mathematics, especially algebra, has been deemed as a gatekeeper that prevents students from accessing greater opportunities (Stinson, 2004). Stinson argues for the reframing of how Algebra is taught and how inclusive teaching and mathematics practices can be implemented by re-situating mathematics into contexts that are relatable to students so they can apply mathematical practices that help them uncover inequities and opportunities that affect them and their communities, so the whole process can become an *empowering mathematics* (Freire, 2000).

Regarding the integration of computer programming with Algebra, the National Council of Teachers of Mathematics (NCTM) has urged students to have regular access to technologies that advance their mathematical sense making, reasoning, problem solving, and communication; and in turn stimulate students' interest and proficiency in mathematics (NCTM, 2023). Even though there is support to access technologies in mathematics classrooms, the possible integration of computer programming and algebra has been debated. While programming languages originated from algebra, each of these registers values, uses, and deals with quantities analytically different. It is argued that, even though the semiotic representations of programming languages and algebraic notation are similar, their meanings differ (Bråting & Kilhamn, 2021; Schanzer, 2017). For example, a variable in programming indicating a process may change value along the execution of a program, while in algebra a variable varies when it describes a relation. Variables have different roles in each system of representations such as controlling the process and storing data in programming and expressing relations in algebra (Bråting & Kilhamn, 2021). Simply, the semantics and syntax of programming are incompatible with mathematics (Schanzer, 2017). For learning's sake, these differences must be taken into account, especially since students will need to move in between registers with overlapping and specific meanings. Bråting and Kilhamn (2021) have urged revisiting definitions to prevent loss of algebraic notation meaning in favor of programming languages.

Nevertheless, contemporary early algebra is blurring the differences between programming and algebra (Bråting & Kilhamn, 2021; Kaput, 2008; Radford, 2018). Authentic integration of mathematics and programming has been deemed possible under three parameters: *tools* (as language itself must enforce basic mathematical concepts), *curriculum* (aligned to national and/or state standards for *mathematics*), and *pedagogy* (focus on great pedagogical teaching techniques rather than just having a great curriculum (Schanzer, 2017).

There are examples of successful integration of programming and algebra (Choppella et al., 2012; Kaufmann & Stenseth, 2020; Nicaud et al., 2003). These examples provide evidence that programming can contribute to greater student motivation for mathematics (Barak & Assal, 2018; Kaufmann & Stenseth, 2020; Ke, 2014; Lambic, 2011; Leonard et al., 2016). For example, Choppella and colleagues (2012) developed a curricular approach that allowed students in the Indian school engineering college to connect high school algebra to the fundamentals of computing through functional programming. This approach was easier to master than traditional imperative programming. Furthermore, Nicaud and colleagues (2003) implemented a computer system (Aplusix) that helped students learn computer algebra and make calculations. The high-school students used commands like Reduce, Expand to ask the system to make some particular calculations such as expand or factor polynomial expressions, solve equations, inequalities, or systems of linear equations. The system provided feedback to students about the steps, so students could modify their strategies. Overall, the system supported the learning of algebra. Teachers argued that the system is a tool that reinforces the acquired knowledge, corrects invalid knowledge, and helps students gain autonomy. The teachers did not need to validate the solutions. Due to this autonomy, students could use the system freely at school or at home.

Also working on this integration, Kaufmann and Stenseth (2020) reported their work with a group of three students learning to program visual effects using Java. The analysis focused on student development of arguments (Lavy, 2006) during mathematics problem solving. The study revealed a progression in student argumentation, from the simplified to the elaborated. For example, in the beginning, the group changed the variables by trial and error, to observe what occurred, and used basic arguments. After a while, the group carefully reread and changed the code as needed by reconsidering their reasoning and their attention on how to solve the problem. It became evident that how the group understands the problem affects the process of how they analyze and explore the solutions and develop different levels of arguments. Kaufmann and Stenseth (2020) argued that it is natural and sensible that the group initially run some intuitive iterations and basic arguments to understand how the program works. The type of task may also have an impact on how students were able to present detailed and mathematically advanced arguments. On the other hand, when students used the trial-and-error methods, it seemed to have had a negative impact on students' mathematical reasoning. The authors claimed that this method lowered the threshold to try new things. The authors concluded that the integration of programming into mathematics education is possible, but a dual perspective is necessary for the teacher to provide advice on mathematics and programming during the process. They recommended more research to observe how students and teachers influence the process of problem solving and their reflection and predictions.

THE *ADVANCING OUT-OF-SCHOOL LEARNING IN MATHEMATICS AND ENGINEERING* CONTEXT: THE AOLME CURRICULUM

Advancing Out-of-School Learning in Mathematics and Engineering (AOLME) provided a learning environment where students experimented with ideas to solidify their understanding of computer programming and mathematics (CPM) practices. This work was done by having middle school students experience two levels of an integrated CPM curriculum using asset-based approaches that honored students' use of their full linguistic repertoire (i.e., Spanish, English, and translanguaging) and their cultural backgrounds to make sense of mathematics and computer programming. Level 1 focused on the basics of computer programming using Python, a text-based computer programming language, to understand how images and video are created down to the pixel level. Level 2 covered object-oriented programming and robotics (see https://aolme.unm.edu/Website-Model/template/index.html). Each level of the curriculum required it be implemented with at least 12 sessions, each lasting 1.5 hours. Seven to eight sessions were devoted to learning about the content, and four or five sessions to working on a final project that consisted of creating a video clip frame-by-frame by specifying colors of rectangular groups of pixels in each frame. The students created video clips based on their personal interests. For the purposes of this study, we focus on Level 1's Session 2: Introduction to Python Programming, in which students explored algebraic concepts such as variables and expressions and formed conjectures through a Number Guessing Game.

METHODS

School Sites

The after-school program was implemented at two bilingual middle schools in an urban and rural context in the Southwest region of the United States. Both schools enrolled primarily Latina/é/o/x students. All students at both schools were on free or reduced-price lunches. Both middle schools offered a bilingual program, which meant that students were taught mathematics in Spanish during sixth and eighth grades, and in English in seventh grade.

Participants: Middle School Students, Co-Facilitators, and Facilitators

AOLME served a total of 135 middle school students who were primarily Latina/é/o/x. Of these students, 24 served as co-facilitators. Being a co-facilitator meant that students experienced at least the Level 1 curriculum and that they attended four Saturday professional development (PD) sessions that lasted approximately 6 hours each. The PD sessions focused on: (a) teaching using an asset-based approach lens that honored the students' bilingualism and different stages of language development of English and Spanish; (b) cooperative learning; (c) mathematics talk moves (Chapin et al., 2022); and (d) the content of the curriculum, allowing the co-facilitators to experience the same tasks that their students would learn. The middle school students worked in small groups that typically consisted of one undergraduate or graduate engineering student facilitator, one co-facilitator who was a middle school student co-teaching with the facilitator, and 2 to 4 middle school students.

For purposes of this study, Fu served as an undergraduate engineering student facilitator. She is bilingual in Vietnamese and English, and she is learning Spanish to communicate with the students. Juanita served as the co-facilitator at the rural middle school, and Melly was the co-facilitator at the urban school. In her self-ratings on Spanish and English proficiency in all domains (speaking, listening, writing, and reading), Juanita rated herself as intermediate in Spanish and advanced in English. Melly rated herself as advanced in both English and Spanish. Melly and Juanita co-taught the curriculum with Fu at the urban and the rural school, respectively. In addition, Fu and Melly worked with two students, Mirasol and Cristie. Juanita worked in a small group which included Vicente, Tina, and Josephina (see Tables 1.1 & 1.2).

TABLE 1.1 Rural School Group of Participants and Their Language Backgrounds

Name	Role	Language 1	Language 2
Fu	Facilitator	Vietnamese	English (Learning Spanish)
Juanita	Co-facilitator	English Advanced	Spanish Intermediate
Vicente	Student	Spanish Advanced	English Beginning
Tina	Student	Spanish Advanced	English Beginning
Josephina	Student	Spanish Advanced	English Beginning

TABLE 1.2 Urban School Group of Participants and Their Language Backgrounds

Name	Role	Language 1	Language 2
Fu	Facilitator	Vietnamese	English (Learning Spanish)
Melly	Co-facilitator	Spanish Advanced	English Advanced
Mirasol	Student	Spanish Intermediate	English Intermediate
Cristie	Student	English Advanced	Spanish Intermediate

Research Design, Data Collection, and Analysis

The research design of this study is a descriptive single case study with embedded units (Yin, 2018). The two small groups were selected because they had a common undergraduate student facilitator who did not share the students' home language; however, Fu was making an attempt to learn Spanish. Also, the co-facilitators were middle school students who self-rated themselves as being intermediate to advanced in their Spanish and English proficiency. These circumstances created a dynamic translanguaging space.

Data sources included video recordings from Session 2 of each group, each video that lasted 1 1/2 hours for a total of 3 hours. The video recordings captured the small group interactions and debriefings at the end of each session with the facilitator and co-facilitator reflecting on their experiences teaching the topic for the day. Another data source was the screen recordings, which focused on capturing the students' coding activity during the group interactions. Student exit interviews (20 to 30 minutes each) were conducted at the end of the program to learn about students' experiences with the program, as well as their self-ratings (on a scale from 1 to 10) on their enjoyment and knowledge of mathematics and computer programming before and after the program. Artifacts such as registration forms, where students rated their Spanish and English proficiency levels in each domain, and students' final projects and their journals were also collected.

Data analysis was conducted using two coding cycles (Saldaña, 2021). The first cycle involved the research team members watching video recordings from the two groups to identify potential moments of interest related to participants' use of language, algebraic thinking, and prediction through open coding. In particular, for the second cycle of coding, we selected the videos using the following criteria: (a) the videos related to Session 2: Programming With Python, as this session included many more algebraic concepts that students were learning initially; (b) the participants used English, Spanish, or translanguaging; and (c) the audio recordings were of good quality. We triangulated the video recordings with other data sources including transcripts, which were completed by a bilingual transcriber and confirmed by research team members, and videos of the monitor screen to check for accuracy. The identified themes from the data during the second cycle of analysis included Making Meaning of Variables Through Playing and Translanguaging and Making Meaning of Variables by Experimenting How to "Break" the Computer.

FINDINGS

To illustrate how the use of language (Spanish/English) mediated middle school Latinx bilingual students' and (co)facilitators' explorations of algebraic thinking through computer programming, we draw from one example from each school in Vignettes 1 and 2.

Vignette 1: Making meaning of Variables Through Playing and Translanguaging

The first vignette comes from Fu working with a group of students (see Table 1.1) who had just joined the group and were barely starting to learn English. Fu collaborated with Juanita, a middle school co-facilitator who was bilingual and felt more fluent in English. Fu, at the end of the lesson, during her reflection that she shared out loud to the camera, she recognized that translanguaging and the promotion of an informal atmosphere were important in the process of supporting the new students.

> Fu (college-student facilitator): Today we got two new students and I was struggling a lot because I don't speak Spanish very much, but Juanita helped out a lot, and we actually had fun. It seems they learned a lot, they understood things, after Juanita explained it to them, it was so good!

Vignette 1 presents how Fu and Juanita, while maintaining an informal environment where computers were part of the thinking team by providing outputs to the input that students provided, also generated curiosity and expectations through the process of asking students to make predictions of what the computer would do. Specifically, the question "*¿Qué va a pasar?*" (What's going to happen?) promoted a context where observing, conjecturing, and generating a pattern and a related hypothesis are supported. Nevertheless, it could also turn into a random guessing game. In either case, the process of making predictions creates a group expectation to check on whether the named prediction is correct or not. It is in this situation where a socialized, collective space is generated. In the first vignette, we describe a group of students exploring algebraic thinking through the learning of basic coding by using the command 'print' when implementing variables in their code that were linked to strings or number operations.

The task of printing out each other's names using strings took place in an informal context of relating to one another what the group developed. As the code developed by a student was ready to be run, the co-facilitators (Juanita and Fu) posed the following question:

Juanita: (middle school co-facilitator) *¿Qué creen que va a pasar?* [quickly reacts to Vicente] *No lo corras todavía. ¿Qué creen que va a pasar?* (What do you think is going to happen? Don't run it yet. What do you think is going to happen?)
Vicente: *Mmm, se cambia el número.* (Mmm, the number changes.)
Tina: *Va aparecer, ¿se pone el nombre?* (It's going to appear. Does the name appear?)
Vicente: *No porque no están las comillas* [makes quotation marks with his fingers]. (No, because they are not in quotation marks.)

After Vicente ran the code, an "error" message was displayed. The group reacted with laughter and different comments, such as "Oh, you're an error, Vicente!" or "This computer is wrong!" and Fu intervened as follows:

Fu: (college student facilitator) Can you translate this for me real quick? [talking to Juanita] Let me write this down. So, in the beginning, we assigned this value [Vicente], to this [variable].
Juanita: *So al principio asignamos el nombre de Vicente para que se diga nombre, ¿okay?* (So, in the beginning we assigned Vincent's name so that it shows the name.)
Fu: So, whenever they say "nombre," they [computer] know that the value is Vicente.
Juanita: *So cuando le digan a la computadora que diga el nombre, la computadora va a pensar en Vicente.* (So, when you tell the computer to say [print] the name, the computer is going to think of Vicente.)
Tina: [nods head] ahh okay
Fu: So, when we say print "nombre," it is not going to print "nombre," they're gonna look for the value of "nombre."
Juanita: *So cuando pongan "nombre" en print, no va a buscar por el nombre [nombre de la variable] va a encontrar la variedad [contenido de la variable].* (So, when you put "nombre" in print, it is not going to look for the name [name of variable], it's going to find the [content of the] variable.)
Fu: "Nombre" equals what?

Juanita: *"¿Nombre" igual a qué?* ("Nombre" is equal to what?)
Vicente: *A uno* (one)
Tina: *Error no igual a Vicente.* (Error not the same as Vincent.)

The group solved the error by linking the determination of a variable with its content; in this case with "string" (text, name) variables. The group noticed how by adding quotation marks on the sides of the word Vicente, it becomes red text, indicating that the content is recognized as a string variable. They discussed how the computer gets to see and read the names when the quotation marks are included and the text turns color red. Each student was able to print out their name. Then, this syntax convention of coding was further elaborated by implementing numerical variables (see Figure 1.1). The group explored the difference between implementing strings and numerical variables by taking turns creating operations, typing them, and "playing" at making them strings or operating on them. In the process of printing out the different numerical operations, students were making predictions of what would happen as the excerpt below presents:

Fu: That's why it doesn't give you six. It only prints three times two.
Juanita: *Por eso no te va a dar seis, solo lo que escribiste.* (That is why it is not going to give you six, only what you wrote.)
Vicente: *Nada más quitar las comillas y te da el resultado.* (Just take away the quotation marks and it gives you the output.)
Fu: But if I delete this [quotation marks], then what do you think is gonna happen?
Vicente: *Seis porque es el resultado.* (Six because that is the answer.)

In Figure 1.1 of the monitor the group was using, it is evident that the variable "name" equaled "3 * 2." It was in quotation marks as a string value, so Fu asked the group, "What will happen if I do this?" and she changed the variable to "3 + 2," and students argued that nothing will change with the numbers because the text still had the quotation marks. By "playing" at printing strings and operating numerical variables and making and discussing predictions of *"qué va a pasar?"* students were able to realize syntax conventions to coding with string variables versus numeric variables (Vygotsky, 1986). In this process, they seamlessly used and created variables. The use of variables was essential to the process of "playing" at printing. Through social play and making predictions, Vicente, Tina, and Josephina had the opportunity to test their ideas and confirm them. Lim and colleagues (2010) argue, "When students predict, as opposed to meticulously working through the steps, they are psychologically relieved from the need for precision and certitude. By temporarily disregarding details, students can focus

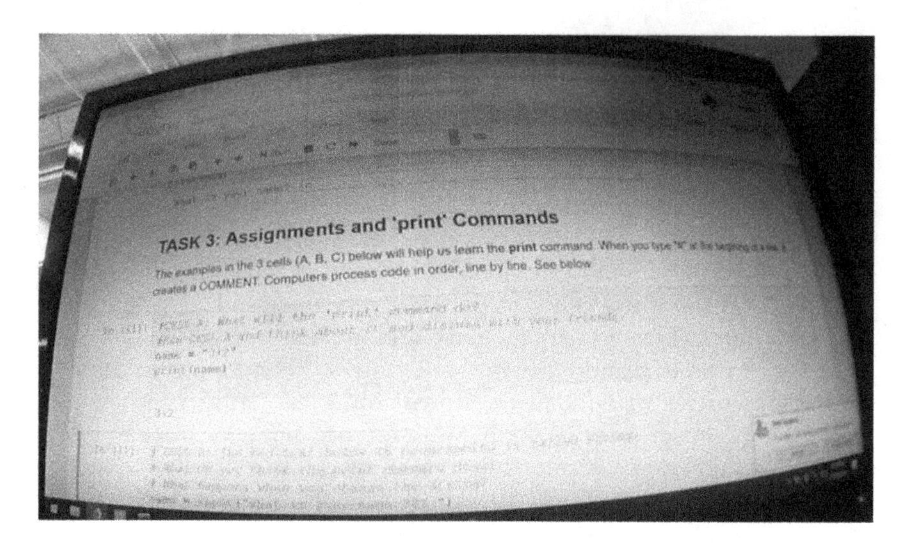

Figure 1.1 "Print" command showing the use of string variables enclosed in quotes as opposed to using numeric variables that do not use quotes.

on essential features and structures" (pp. 597–598). This informal process helped them focus on specific aspects of what makes a string value vs a numerical value and, based on that, run the program and see what happened (*¿Qué pasó?*). This helped them not be intimidated about being right or wrong, but to check if their predictions and their friends' indicated what was going to happen. Once they corroborated and discussed the results of their predictions after running the code, they applied some patterns linked to print commands and code syntax conventions, such as string values being in quotation marks and being in red text that they had used. Then, they moved to a generalized way of using these concepts (Lim et al., 2010). This realization came about by checking their predictions and seeing if a pattern they discovered would hold true for "all" cases. Then, with the support of Fu and Juanita, students formulated general statements and conjectures and tested how predicted, generalized patterns in strings and operation of numbers worked using different variables that they co-constructed based on previous "experiments" (Choppella et al., 2012).

Vignette 2: Making Meaning of Variables by Experimenting How to "Break" the Computer

Similar to the first vignette, the second vignette involves Fu collaborating with a small group of students who were all at intermediate or advanced levels of Spanish and English language development (see Table 1.2). Fu worked with Melly, a middle school co-facilitator who was bilingual and rated herself as being advanced in Spanish and English. At the end of the session, Fu debriefed on what they accomplished as a group, reflecting primarily on what the students had learned and what the experience was like in an environment that promoted a hybrid space to generate curiosity of algebraic concepts through the use of games.

> **Fu:** (college student facilitator) Okay today they really had a lot of fun in my opinion because they really liked the games and I think they actually learned what happened with the commands, string, print, and everything so yeah it was good.

One of the tasks that students engaged with during Session 2: Programming With Python was about programming something fun with their groups (see Figure 1.2). Also, similar to Vignette 1, this task involved a process of predicting what would be printed when using the print command and experimenting with operations on numbers and strings, so that students could establish a relationship between the input and output of the variables that they were defining and printing.

The group decided to start experimenting by programming number operations. Taking turns, the students tried different operations. Melly created a variable that they called "y" and, as portrayed in Figure 1.2, students started

Figure 1.2 Task 4: Program something fun in your group.

creating an equation inputting symbols and long numbers. The students' idea of using big numbers was to "break" the computer. As students were ready to run the code, Fu asked students to make a prediction. In the excerpt below students provided answers to Fu's prompt on predicting what they thought the computer was going to print (*When you print the variable what's going to show?*), a question similar to "*¿Qué va a pasar?*" posed in Vignette 1.

Cristie: The letter y.
Melly: It won't work. It won't work.
Fu: It won't work? What do you think it's gonna do?
Melly: It's gonna print the whole thing. It's gonna print the equation.
Mirasol: Umm...
Fu: The whole thing? Like exactly as it looks here? [pointing numbers in equation]
Mirasol: Yeah.
Melly: The answer.
Fu: Which one?
Melly: The answer to the whole thing.

After everyone had presented their predictions, Melly ran the code and each corroborated their conjectures. Fu asked them to reflect on what had happened and why; as presented below.

Fu: What did it [the code] do guys?
Cristie: The answer.
Mirasol: It gave the answer!
Fu: Okay so what... [To Melly] You were right.
Melly: Yes, I'm always right.
Fu: [To Mirasol] Why did it not print the whole thing? It only prints the whole thing if you have what?
Mirasol: Umm.
Cristie: Input.
Melly: Quotes.
Mirasol: Quotes.
Fu: It only prints...
[Students laugh.]
Fu: It only prints exactly if you have it in quotation right?
Mirasol: Yeah.
Fu: [To Cristie] And then why doesn't it print y?
Mirasol: Because I think, because of quotations.
Fu: Yeah! There's no quotations around y right? So if you want to print the letter y you have to add quotations here [in the

parentheses of the print command], right? When there's no quotation what does it look for?

Cristie: The answer.

Fu: It looks for the variable yeah. It looks for whatever is here. So it looks for the variable y. Is there a variable y? Here right? So it sets Y equal to this whole thing. It's gonna find the answer to this and it didn't break. You guys didn't try, you guys didn't break... it gave you the answer.

Merari: What!

Cristie: We have to do it again!

The students together with Fu reviewed each of their predictions considering what would have been needed to make their predictions come about. Students had previously explored strings using quotes and through this "experimentation" students revisited something they had already talked about, but the interaction brought it back. Through the process of students differentiating how their predictions would play out in the printing of variable "y" students were able to engage in the process of proving and disproving conjectures. This process helped them reinforce what they had practiced earlier. Through their conjectures, the students developed arguments in the application of what they had just learned. These arguments all revolved around how a variable was defined and printed. As Lim and colleagues (2010) have argued, predicting includes the testing of conjectured patterns. This testing process inevitably supported the construction of knowledge. As such, the whole process in Task 4 helped students revisit ideas they chose to experiment with. Moreover, the girls did not give up and persevered in finding a code that could "break" the computer. Their attempts continued, including big numbers to fulfill this goal. As a result, Fu, using playful talk, pretended to bet and invited them to note the operation she created. Fu asked: "Is it gonna break? Is this math right? Should it give me an answer? I bet you it's gonna break the computer. Run it!" As Melly ran the code, students were surprised that the computer could not process the division of a number by zero. Fu sparked the girls' curiosity and engaged them in thinking of the mathematics in the operation by encouraging them to make conjectures about why a number cannot be divided by zero:

Fu: Why? Why can't it [the computer] find the answer for 1 divided by zero?

Melly: Because there's no possible way.

Melly: Because there is no division that can divide by zero.

Fu: Uh huh. You cannot divide by zero right?

Melly: No.

Fu: [looking at Mirasol] Can you divide by zero?

Mirasol: No.

With this problem in mind, students explored with Fu examples of how one can divide by 0. This interaction included thinking of situations and dividing objects such as a pie. The situation was thinking about having zero pies and giving them to three people and asking them how many pies each person gets. Some students were still not convinced about this conjecture, so Fu continued to provide a space where students could try different entries and run the code by operating with bigger numbers, using the print command, dividing by zero, and trying to break the computer. In the end, all students conjectured that no number can be divided by zero. What is important is that students participated in an activity that involved a playful way of predicting what would be displayed through the print command. Through curiosity and spontaneous exploration (García & Wei, 2014; Vygotsky, 1986), students engaged in a process of learning mathematics and computer programming. They engaged in a trial and error process in the attempt to "break" the computer. In contrast to previous studies (Kaufmann & Stenseth, 2020), the trial-and-error process that students engaged in attempting to break the computer generated more interest in generating codes that could do that. However, it is also important to note that in the trial-and-error process, students were supported by the facilitators. This support helped students observe in more detail what occurred especially when they made predictions, developed arguments, and discussed their implications by running and "playing" (Vygotsky, 1986) and replaying the codes they created. The group carefully reread and changed the code as needed by reconsidering their reasoning and their attention on how to solve the problem or task at hand (Bråting & Kilhamn, 2021; Kaufmann & Stenseth, 2020; Schanzer, 2017). Moreover, in all this process students created and used variables as an essential part of their discussions. These variables were linked to both mathematical and programming variables. Students were able to develop "fluency" in the functional use and implementation of variables (Choppella et al., 2012). Students in general reported feeling motivated in doing mathematics this way and learning about computer programming (Barak & Assal, 2018; Kaufmann & Stenseth, 2020; Ke, 2014; Lambic, 2011; Leonard et al., 2016).

CONCLUSIONS AND IMPLICATIONS

The process of combining mathematics with computer programming supported student learning fundamental programming concepts, such as defining a variable, printing, and running the code that students were developing. Before running a code, students were asked to think "*¿Qué va a pasar?*" in order for them to predict the possible outcomes. The process of predicting "activate[s] and refine[s] their[students'] existing knowledge" (Lim et al.,

2010, p. 598). This process helped uncover students' conceptions on concepts such as a variable and how it is defined in both algebra and computing. Students seamlessly used and created variables in the process of printing variables associated with strings and numerical values. As students "played" at conjecturing what will happen or be printed with the code that some had created, students paid attention to issues of coding syntax conventions (e.g., using quotation marks) using strings and numbers (Vygotsky, 1986).

The process of corroborating and discussing the results of their predictions, after running the code, supported them in identifying patterns that in turn helped them identify basic features of printing commands, strings, defining and operations using numerical variables, and creating equations. It is essential to note that the group in the first vignette was able to engage in all these processes in a "playful" social context where the use of Spanish and English also were seamlessly integrated, a "translanguaging" context (García & Wei, 2014) and procedures where the means of expression included predicting, thinking, and laughing about coding and operating diverse forms of variables.

Similarly in the second vignette, students learned about the connections between mathematics and computer programming. Specifically, they learned how variables worked by printing the variable y and proving or disproving their conjectures. The facilitators and the students engaged in spontaneous exploration (García & Wei, 2014; Vygotsky, 1986) where they learned mathematics and computer programming through a trial-and-error process in an attempt to "break" the computer. In addition, the students and the facilitator learned while playing by placing bets on their predictions. This provided a playful way to learn about variables, strings, and the print command. Students also learned about making conjectures, specifically that no number can be divided by zero.

Findings from this study have important implications for mathematics classrooms in which Spanish and English are used as a medium of instruction. From our work, it is evident that a third space provided bilingual learners a space to demonstrate their brilliance in learning mathematics and programming concepts. As Lim et al. (2010) recommend, "a teacher should envision how the activity will unfold, anticipate students' predictions and be prepared to respond appropriately" (p. 606). Teachers can include activities that engage students in sharing reasoning about their predictions and resolving differences between what students predicted and their solutions. In doing so, teachers can generate discussions among students to address cognitive dissonance and direct students' attention towards a certain mathematical or computing concept.

Based on our findings, we are advocating for a teaching and learning of computer programming and mathematics that goes beyond the goal of "access" to technology and STEM (science, technology, engineering, and mathematics) practices for BIPOC (Black, indigenous, and people of color) students. Access alone is insufficient; increasing the number of "BIPOC bodies in computing majors or in computing jobs will not produce racial justice" (Shah & Yadav, 2023, p. 472). The reality is that "computer science (CS) for all" is not a jobs program for BIPOC communities at scale. CS pedagogies are often aligned with core school expectations, cookie-cutter curricula that put students in passive roles, rather than active participants. Shah and Yadav (2023) suggest a culturally responsive computing (CRC) approach. This approach empowers students "as leaders of their own creativity and learning" (Reich, 2020, p. 91) and connects computing to BIPOC students' families, communities, interests, and lived experiences as a "vehicle for learning" (Ladson-Billings, 1995, p. 161). The authors assert that racial equity in computing education means that BIPOC students participate in authentic programming learning experiences and use computing to pursue their own interests and express their personal agency, so their own backgrounds and experiences are leveraged. In our work (see LópezLeiva et al., 2022), we have tried to include a CRC approach through the process of promoting a teaching and learning environment that is inclusive of students' linguistic repertoires, their way of thinking and speaking by starting where they are through the use of translanguaging, spontaneous learning, informal interactions, and activities that support student predictions so that students can present and build upon their current knowledge. The goal is that students may figure that computer programming is a cultural practice of their world that they can use as a literacy practice to express ideas and generate new ones. As Juanita mentioned at the beginning, bilingual learners "are smart learners." What we learned from our work is that bilingual learners can engage deeply in algebraic thinking and computer programming by testing their ideas and confirming them through play and prediction.

REFERENCES

Barak, M., & Assal, M. (2018). Robotics and STEM learning: Students' achievements in assignments according to the P3 task taxonomy—practice, problem solving, and projects. *International Journal of Technology and Design Education, 28*(1), 121–144. https://doi.org/10.1007/s10798-016-9385-9

Bråting, K., & Kilhamn, C. (2021). Exploring the intersection of algebraic and computational thinking. *Mathematical Thinking and Learning, 23*(2), 170–185. https://doi.org/10.1080/10986065.2020.1779012

Celedón-Pattichis, S., Kussainova, G., LópezLeiva, C., & Pattichis, M. S. (2022). "Fake it until you make it": Participation and positioning of a bilingual Latina student in mathematics and computing. *Teachers College Record, 124*(5), 186–205.

Chapin, S., O'Connor, C., & Anderson, N. C. (2022). *Talk moves: A teacher's guide to using classroom discussions in math, grades K–6.* Heinemann.

Choppella, V., Kumar, H., Manjula, P., & Viswanath, K. (2012). From high-school algebra to computing through functional programming. *IEEE Fourth International Conference on Technology for Education*, Hyderabad, India, pp. 180–183, https://doi.org/10.1109/T4E.2012.42

Code.org Advocacy Coalition & CSTA. (2018). *2018 state of computer science education: Policy and implementation.* https://advocacy.code.org/2018_state_of_cs.pdf

Freire, P. (2000). *Pedagogy of the oppressed (30th anniversary ed.).* Continuum.

Garcia, O., & Wei, L. (2014). *Translanguaging: Language, bilingualism, and education.* Palgrave McMillan.

Gutiérrez, K., Baquedano-López, P., & Tejeda, C. (1999). Rethinking diversity: Hybridity and hybrid language practices in the third space. *Mind, Culture, and Activity, 6*(4), 286–303.

Jacob, S. R., Montoya, J., & Warschauer, M. (2022). Exploring the intersectional development of computer science identities in young Latinas. *Teachers College Record, 124*(5), 166–185.

Kaput, J. (2008). What is algebra? What is algebraic reasoning? In J. Kaput, D. Carraher, & M. Blanton (Eds.), *Algebra in the early grades* (pp. 5–18). Lawrence Erlbaum.

Kaufmann, O. T., & Stenseth, B. (2020). Programming in mathematics education. *International Journal of Mathematical Education in Science and Technology.* https://doi.org/10.1080/0020739X.2020.1736349

Ke, F. (2014). An implementation of design-based learning through creating educational computer games: A case study on mathematics learning during design and computing. *Computers and Education, 73*, 26–39. https://doi.org/10.1016/j.compedu.2013.12.010

Ladson-Billings, G. (1995). But that's just good teaching! The case for culturally relevant pedagogy. *Theory Into Practice, 34*(3), 159–165.

Lambic, D. (2011). Presenting practical application of mathematics by the use of programming software with easily available visual components. *Teaching Mathematics and Its Applications, 30*(1), 10–18. https://doi.org/10.1093/teamat/hrq014

Lavy, I. (2006). A case study of different types of arguments emerging from explorations in an interactive computerized environment. *The Journal of Mathematical Behavior, 25*(2), 153–169.

Leonard, J., Buss, A., Gamboa, R., Mitchell, M., Fashola, O. S., Hubert, T., & Almughyirah, S. (2016). Using robotics and game design to enhance children's self-efficacy, STEM attitudes, and computational thinking skills. *Journal of Science Education and Technology, 25*(6), 860–876. https://doi.org/10.1007/s10956-016-9628-2

Lim, K. H., Buendía, G., & Kim, O. K., Cordero, F., & Kashmir, L. (2010). The role of prediction in the teaching and learning of mathematics. *International Journal of Mathematical Education in Science and Technology, 41*(5), 595–608.

LópezLeiva, C., Noriega, G., Celedón-Pattichis, S., & Pattichis, M. S. (2022). From students to cofacilitators: Latinx students' experiences in mathematics and computer programming. *Teachers College Record, 124*(5), 146–165.

Murphy, C. (2022). *Vygotsky and science education.* Springer Nature Switzerland.

National Council of Teachers of Mathematics. (2023). *Equitable integration of technology for mathematics learning.* https://www.nctm.org/Standards-and-Positions/

Position-Statements/Equitable-Integration-of-Technology-for-Mathematics-Learning/

Nicaud, J. F., Bouhineau, D., Chaachoua, H., Huguet, T., & Bronner, A. (2003). A computer program for the learning of algebra: Description and first experiment. *PEG 2003 Conference,* St. Petersburg, Russia, (p. 7). hal-00190396

Pozos, R. K., Severance, S., Denner, J., & Téllez, K. (2022). Exploring design principles in computational thinking instruction for multilingual learners. *Teachers College Record, 124*(5), 127–145.

Radford, L. (2018). The emergence of symbolic algebraic thinking in primary school. In C. Kieran (Ed.), *Teaching and learning algebraic thinking with 5- to 12-year-olds* (pp. 3–25). Springer.

Radke, S. C., Vogel, S. E., Ma, J. Y., Hoadley, C., & Ascenzi-Moreno, L. (2022). Emergent bilingual middle schoolers' syncretic reasoning in statistical modeling. *Teachers College Record, 124*(5), 206–228.

Reich, J. (2020). *Failure to disrupt: Why technology alone can't transform education.* Harvard University Press.

Saldaña, J. (2021). *The coding manual for qualitative researchers.* SAGE.

Schanzer, E. (2017). *Integrating computer science in math: The potential is great, but so are the risks.* AMS, Blog on Teaching & Learning Mathematics. Retrieved from https://blogs.ams.org/matheducation/2017/01/09/integrating-computer-science-in-math-the-potential-is-great-but-so-are-the-risks/

Shah, N., & Yadav, A. (2023). Racial justice amidst the dangers of computing creep: A dialogue. *TechTrends, 67,* 467–474. https://doi.org/10.1007/s11528-023-00835-z

Stinson, D. W. (2004). Mathematics as "gate-keeper" (?): Three theoretical perspectives that aim toward empowering all children with a key to the gate. *The Mathematics Educator, 14*(1), 8–18.

Vogel, S., Hoadley, C., Castillo, A. R., & Ascenzi-Moreno, L. (2020). Languages, literacies and literate programming: Can we use the latest theories on how bilingual people learn to help us teach computational literacies? *Computer Science Education, 30*(4), 420–443. https://doi.org/10.1080/08993408.2020.1751525

Vygotsky, L. (1978). *Mind in society: The development of higher psychological processes.* Harvard University Press.

Vygotsky, L. (1986). *Thought and language.* MIT Press.

Yin, R. K. (2018). *Case study research and applications: Design and methods* (6th ed.). SAGE.

Zong, J. (2022). *A mosaic, not a monolith: A profile of the U.S. Latino population, 2000–2020.* UCLA Latino Policy and Politics Institute. https://latino.ucla.edu/research/latino-population-2000-2020/#:~:text=Between%202000%20and%202020%2C%20Latinos,than%20Latinos%20over%20this%20period

CHAPTER 2

WRITING IN SUPPORT OF MATHEMATICAL UNDERSTANDING FOR EMERGENT BILINGUAL LEARNERS

Marco A. Bravo
Santa Clara University

Maria Valencia-Orozco
Santa Clara University

ABSTRACT

This chapter examines the possibilities of integrating writing and mathematics for Emergent Bilingual Learners (EBLs). Elementary-grade Spanish/English bilingual students ($n = 55$) responded to a writing prompt at the onset and end of an academic year that elicited their explanation as to how they solved a mathematical problem. Students could select whether to provide their response in English or Spanish. The responses were subsequently assessed with a rubric, gauging EBLs' abilities to provide their mathematical reasoning in their expla-

Mathematics Instruction in Dual Language Classrooms, pages 25–39

nation, how their mathematical computations leveraged appropriate concepts and procedures, how they utilized mathematical vocabulary, and utilized and included appropriate visual literacy (e.g., number line, graph, illustration). The quantitative analysis (paired t-test) from these data show promise for synergistic effects on both writing and mathematical knowledge when these disciplines are integrated. Study results illuminate the need for additional opportunities for writing to have a role in supporting EBLs' mathematical learning.

With increasing numbers of Emergent Bilingual Learners (EBLs) across the country, exploring how best to deliver instruction to this population and trace their development has gained considerable importance. EBLs, also referred to as English learners, constitute the fastest growing sector of the school age population (National Center of Educational Statistics, 2023). Children and youth of school age who spoke a language other than English at home represented 10.4% of the total school enrollment (National Center for Education Statistics, 2023), the majority of whom are non-European, non-English speaking, and extremely diverse (U.S. Government Accountability Office, 2022). This diversity presents a unique paradox for schools. On the one hand, such diversity bestows a challenge for teachers attempting to accommodate multiple languages and cultures in the classroom. Yet, this same "dilemma" affords all students access to multiple perspectives and diverse approaches to problem-solving in school (Paris, 2012).

In an analysis of effective schools that served large numbers of young bilinguals, Garcia (1999) found of utmost importance high expectations, including but not limited to bilingualism and biliteracy goals, as is offered by dual language programs, along with on-going effective staff development. Successful schools implemented: (a) maintenance bilingual program, accomplishing mature English literacy through primary language instruction; (b.) demanding grade level content instruction instead of a watered-down version of the curriculum; (c) instruction organized in innovative ways, where instructional strategies scaffold student learning (e.g., specially designed academic instruction in English-SDAIE, sheltered English); (d) instructional programs which extended instructional time (e.g., after-school programs, voluntary Saturday, summer programs); and (e) parent outreach. These findings have been echoed by others (Basterra, 1998; Baur & Gort, 2012; Cummins, 2000; Valdés, 2005).

Garcia termed these collective practices a *responsive pedagogy* which also encompass practical, contextual, and empirical knowledge and a "world view" of education that evolves through meaningful interactions among teachers, students, and other school community members. These strategies expand students' knowledge beyond their own immediate experiences while using those experiences as a sound foundation for appropriating new knowledge. This chapter details one project's attempt to illustrate the responsive pedagogies implemented by beginning teachers as they integrated writing into their mathematics curriculum in dual language settings.

BUILDING MATHEMATICAL AND BILINGUAL PROFICIENCIES

Language is essential in learning mathematics (Moschkovich, 2013). Shifting instructional reforms call for students to explain, defend, and communicate their mathematical reasoning (National Governors Association, 2010). This need to use language to learn and participate in mathematics activities intensifies for EBLs (Rubinstein-Avila et al., 2015) who may still be developing their bilingual proficiency. Yet, positioning language and literacy as a tool for learning (Dyson, 1993; Moll, 2001), can have synergistic effects on mathematical learning and provide EBLs an opportunity to develop their biliteracy skills. That is, instruction that contextualizes the use of literacy to learn mathematics in an authentic manner, can provide the type of instruction EBLs require as they attempt to learn both content and language simultaneously. Below we share instructional models that attempt to address mathematics learning through the integration of writing instruction into mathematics curriculum.

Providing EBLs with opportunities to write about their mathematical thinking, creates a window for teachers to view EBLs processes in problem-solving and provides a view into their writing development as well. Similar findings have been found with discourse practices in mathematics (Celedón-Pattichis & Turner, 2012). As with discourse, writing in mathematics can have a variety of purposes, including to allow students to explore and make sense of a problem, explain mathematical procedures, build or challenge an argument, and to creatively report original ideas (Casa et al., 2016). Putting writing to work in the service of mathematics has proven to support mathematical learning.

When students write about their mathematical learning, they develop metacognition as they consolidate and reflect on their thinking (Morgan, 1994; Sierpinska, 1998). Pugalee (2004) found students exhibited several metacognitive behaviors when writing about their mathematical problem-solving processes. Using qualitative coding schemes, he noted students' writing provided an opportunity to orient themselves through note taking to better understand the problem. Writing also allowed students to organize their plan to solve the problem before actually solving the problem and finally verifying the problem-solving strategy. This additional layer to explicitly illustrate their problem strategy through writing, sharpened their metacognition which in turn promoted better problem-solving skills (Bicer et al., 2013). For EBLs in particular, this additional opportunity to process their learning through writing is exactly the type of multimodal experience they require to gain access to mathematical learning (Musanti et al., 2009).

Writing is also a great tool for increasing students' knowledge about mathematics (Meel, 1999). Mathematical journals have proven to be academically fruitful for students' mathematics learning. Borasi and Rose (1989) report strong gains in mathematical content knowledge for students that were instructed on how to use their journal to see relationships between prior learning and current learning goals, cement understanding of mathematical concepts, and to recognize and fix their mathematical misunderstandings. Researchers have also noted the use of mathematical journals to write about what students don't understand about the mathematics they are learning (Ashlock, 2006). This type of information can serve as an instructional compass for teachers to further support students. Moreover, allowances for EBLs to write using their full linguistic repertoire affirms EBLs' identity, creating a more inclusive classroom space (Aguirre et al., 2013) which also begets stronger mathematical learning.

While we know these instructional models are effective, fewer studies exist that chronicle how teachers implement writing into mathematics instructional time. Celedón-Pattichis and Turner (2012) provide insight into EBLs' experiences learning the discourse of mathematics, of which they include writing. They share models utilized by teachers that supported very young EBLs use of more precise mathematical language, both orally and in written form. Instruction allowed EBLs in a bilingual context to explore a mathematics word problem, utilize various tools (e.g., counters, cubes) and whiteboards to write their reasoning. Such models move beyond the educative features of curriculum that often suggest mathematical remediation (de Araujo & Smith, 2022). Chval et al. (2021), similarly suggest amplifying the learning opportunities for EBLs. In their review of the literature, strong academic results are possible when EBLs are provided with explicit and systematic instruction on the various genres of mathematical writing. In particular, offering scaffolds in the form of native language use, sentence frames, and graphic organizers, provide the support for EBLs to reap the benefits of leveraging writing to access mathematical learning.

The present study adds to this research base of mathematics and language integration. Utilizing EBLs' writing responses, we capture the possibilities of building both bilingual and mathematical proficiencies when instruction includes explicit opportunities to learn to explain mathematical processes through writing.

METHODS

Study Overview

To gauge the mathematical writing of the EBLs in this study, EBLs' writing samples were collected at the onset and end of an academic year. The classroom teachers of these students were part of a research effort that studied the possibilities and limits of mathematics and language integration in dual language classroom settings.

The present investigation is guided by the following research question and sub-question:

> **RQ1:** *How does mathematical writing mature over time for students who are taught to write in Spanish and English during mathematics instruction?*
>
> **RQ1a:** *Do students show strength in particular domains of mathematical writing (Math Reasoning, Math Computation, Math Vocabulary, Math Literacy)?*

Teachers implemented lessons throughout the academic year integrating vocabulary, discourse, and literacy. Teachers targeted mathematical vocabulary that was both procedural (e.g., quotient, variable) and conceptual (e.g., vertices, associative property) in nature through such strategies as reviewing the morphological structure of words and using cognates to gain access to unfamiliar mathematical words in one language or the other. Teachers also included instructions that demystified the nature of mathematical explanations and argumentation. Lessons illustrated to students the structures of both of these discursive practices, including the need to provide a claim, reasoning, and concluding statement that links the two. EBLs were given freedom to participate in these discussions utilizing their full linguistic repertoire. Similarly, teachers also made evident the nature of mathematical writing during their mathematics instruction. Teachers provided students with opportunities to journal about their mathematics experiences, practice explaining their mathematical thinking through writing with several prompts that students responded to during either mathematics or language arts class.

We test the efficacy of these practices with a pre and post writing assessment.

Administration of the Math Writing Assessment System

Teachers were provided with scripted instructions for the administration of the Math Writing Assessment System (MWAS) and in the respective language in which mathematics was being assessed. The MWAS was administered at the onset, and end of the 2021 academic year. The administration of the assessment was the same at each grade level. Table 2.1 provides the prompts students responded to.

Scoring Process

Three researchers calibrated scores on twelve student papers. Any discrepancies in scoring were discussed, (referencing both the writing rubric and anchor papers) until there was a majority consensus on the scoring of the initial five writing samples. Following and during the scoring of student papers, through what we termed "literacy digs," we talked with each other

TABLE 2.1 Writing Prompts

Language/ Grade	Pre	Post
Spanish K	*Instrucciones:* Usa el espacio debajo para mostrar tu trabajo de matemáticas. En las líneas de abajo, explica tu respuesta. ¿Cuántos estudiantes hay en tu clase? Haz un dibujo que te ayude a resolver el problema. **English Translation:** *Directions: Use the space below to show your math work. On the lines below, explain your answer.* *How many students are in your class? Draw a picture to help you solve the problem.*	*Instrucciones:* Usa el espacio en la hoja para demostrar tu trabajo de matemáticas. En las líneas explica tu respuesta. Carlos tiene 3 pelotas y Ángel tiene 2 pelotas. ¿Cuántas pelotas hay en total? **English Translation:** *Directions: Use the space on the sheet to demonstrate your math work. On the lines explain your answer.* *Carlos has 3 balls and Angel has 2 balls. How many balls do they have?.*
Spanish 2nd	*Instrucciones:* Usa la casilla para mostrar tu trabajo matemático. En las líneas de abajo, explica con palabras tu respuesta. Tu profesora quiere que tú dibujes a tu animal favorito. Tú tienes 33 crayones y 18 marcadores. ¿Cuántos más crayones que marcadores tienes? Muestra tu trabajo matemático en la casilla de abajo. **English Translation:** *Directions: Use the box to show your math work. On the lines below, explain your answer in words.* *Your teacher wants you to draw your favorite animal. You have 33 crayons and 18 markers. How many more crayons than markers do you have? Show your math work in the box below.*	*Instrucciones:* Usa el espacio en la hoja para demostrar tu trabajo de matemáticas. En las líneas explica tu respuesta. El papá de Alex le permitió sacar $25 dólares de la alcancía para gastar en la feria. En la feria, Alex decidió comprar un helado de $12 dólares. ¿Cuántos dólares de más le quedan a Alex para gastar en la feria? Muestra tu trabajo matemático en el espacio de abajo. **English Translation:** *Directions: Use the space on the sheet to demonstrate your math work. On the lines explain your answer.* *Alex's dad let him take $25 out of the piggy bank to spend at the fair. At the fair, Alex decided to buy a $12 ice cream. How many extra dollars does Alex have left to spend at the fair? Show your math work in the space below.*
English 4th	*Instrucciones:* Utilice el cuadro para mostrar su trabajo de matemáticas. En las líneas siguientes, explique su respuesta. Su salón de clases necesita una alfombra nueva. El director le ha pedido que averigüe cuánta alfombra	*Instrucciones:* Utilice el cuadro para mostrar su trabajo de matemáticas. En las líneas siguientes, explique su respuesta. Su maestra de arte quiere instalar un nuevo papel pintado decorativo en el salón. La maestra le ha pedido

(continued)

TABLE 2.1 Writing Prompts (Continued)

Language/ Grade	Pre	Post
	necesitará comprar. El tamaño de su salón de clases es de 35 pies por 52 pies. Escriba una ecuación que represente el problema y luego resuelva el problema. **English Translation:** *Directions: Use the box to show your math work. In the lines below, explain your answer.* *Your classroom needs a new carpet. The principal has asked you to find out how much carpet he will need to buy. The size of your classroom is 35 feet by 52 feet. Write an equation to represent the problem and then solve the problem.*	que averigüe cuánto papel pintado necesitaría comprar. El tamaño de las paredes es de 33 pies por 52 pies. Escriba una ecuación que represente el problema y luego resuelva el problema. **English Translation:** *Directions: Use the box to show your math work. In the lines below, explain your answer.* *Your art school teacher wants to install a new wallpaper decoration in the classroom. Your teacher has asked you to find out how much wallpaper she would need to buy. The size of the wall is 33 feet by 52 feet. Write an equation to represent the problem and then solve the problem.*
English 5th	*Instrucciones:* Utilice el cuadro para mostrar su trabajo de matemáticas. En las líneas siguientes, explique su respuesta. Su clase está haciendo una excursión al parque. Se tarda 45 minutos en caminar hasta el parque. Se necesitan 4⁄5 de ese tiempo para caminar de regreso a la escuela. ¿Cuánto duró el camino de regreso a la escuela? **English Translation:** *Directions: Use the box to show your math work. In the lines below, explain your answer.* *Your class is taking a field trip to the park. It takes 45 minutes to walk to the park. It takes 4⁄5 of that time to walk back to school. How long was the walk back to school?*	*Instrucciones:* Utilice el cuadro para mostrar su trabajo de matemáticas. En las líneas siguientes, explique su respuesta. El entrenador de correr lleva al equipo para practicar en el campo. El equipo necesita 55 minutos para completar su práctica. El equipo utiliza 3⁄5 de ese tiempo caminando por el campo. ¿Cuánto tiempo pasa el equipo caminando por el campo? **English Translation:** *Directions: Use the box to show your math work. In the lines below, explain your answer.* *The cross-country coach takes the team to practice on the field. It takes the team 55 minutes to complete their practice. The team spends 3⁄5 of that time walking around the field. How long does the team spend walking on the field?*

about the trends in writing among the students. Researchers scored math writing samples only after obtaining an 91% inter-rater reliability on 10 student-writing samples. The scoring team was provided with a set of anchor papers that were previously established by the team for reference purposes.

A total of 10% of math writing samples were re-scored to reestablish inter-rater reliability at the mid-point of scoring the math writing responses.

Rubric. The rubric includes four domains of strategies and skills that students were measured against: math reasoning, math computation, math vocabulary, and math literacy. *Math Reasoning* assesses the student's ability to organize and present a logical explanation and justification as well as understanding of the appropriate math concepts to arrive at the correct solution. *Math Computation* focused on the use of the appropriate procedures (skills) to arrive at the correct solution. *Math Vocabulary* emphasized the use of mathematical vocabulary with precision, both conceptual (e.g., vertices, associative property) and procedural (e.g., quotient, variable) terms. *Math Literacy* focused on the use of such visual literacies as graphs, number lines, groupings to explain their mathematical understanding. Each category is divided into several scoring sections that gauges the writing development within each of the four categories in a developmental mode. Every writing sample received a score for each of the domains that range from 1–4. The rubric also includes a benchmark score (3) for each domain based on district standards.

Participants

A total of 56 students from three dual language schools participate in the study (see Table 2.2). Participants were enrolled in kindergarten, 2nd, 4th, and 5th grade. They met specific criteria, including having been present for both MWAS administrations.

These schools enroll a high percentage of English language learners and are designated "high need schools." The elementary schools offered Spanish/English bilingual programs from kindergarten to fifth grade. Literacy and mathematics were taught in both languages with an emphasis on the students' home language.

TABLE 2.2 School Setting

	Rydell Elementary School	Whisper Charter School	Avanzar Elementary School	Avanzar II Elementary School
Program Type	Bilingual Program	Dual Language Program	Dual Language Academy	Dual Language Academy
Students receiving free and reduced lunch	63%	64%	55%	79%
Percent English Language Learner	54%	55%	38%	66%

Data Collection

Student MWAS writing responses were collected in October/November and again in March/April. The upper grade teachers chose to administer the writing prompt in English, while the lower grade teachers (K–2) administered the writing prompt in Spanish. Teachers administered the assessment with guidance from the research team, including making allowances for students to respond to Spanish or English, providing as much time as needed, explanations of the administration in both languages, and allowing the students to complete the task independently.

Data Analysis

The writing samples were scored by the authors only after reaching an inter-rater reliability score of 91%. The student mathematics writings were collected and assessed according to specific criteria specified in the MWAS rubrics. Each individual student paper received four scores (scaled between one and four), each score representing student skills in the four domains of the rubric—mathematical reasoning, mathematical computation, mathematical vocabulary, and mathematical literacy. A domain score is attained by meeting all the criteria within that cell.

A Paired sample *t*-test was performed to compare the pre and post scores of the four domains of the rubric, per grade levels K/2nd and 4th/5th. Any missing values disqualified the participant from the analysis.

RESULTS AND DISCUSSION

Holistically, there was a statistically significant difference between pre ($M = 2.32$, $SD = 0.99$) and post scores ($M = 2.63$, $SD = 1.17$) for 4th/5th grade students but not so for the kindergarten and second grade EBLs. After disaggregating the data by domain, computation (Pre: $M = 2.66$, $SD = 1.19$; Post: $M = 3.29$, $SD = 0.93$) showed statistically significant differences for kindergarten and second grade EBLs and vocabulary (Pre: $M = 1.96$, $SD = 0.91$; Post: $M = 2.46$, $SD = 1.18$) for the 4th/5th grade EBLs. There was no statistically significant difference for the other domains for both grade level groups.

Fourth/Fifth Grade EBLs

The quantitative analysis of these data show promise for synergistic effects on both writing and mathematical knowledge when these disciplines are integrated. When analyzing the data across the four domains (mathematics reasoning, mathematics computation, mathematics vocabulary, & mathematics literacy), the upper grade students demonstrated significant growth from the beginning of the year as compared to the end of the year. They made the strongest gains in the vocabulary domain, increasing from an average of 1.96 ($SD = 0.91$) to 2.46 ($SD = 1.18$) on this rubric domain. Students also demonstrated strong mathematical literacy scores, increasing from 2.63 ($SD = 0.77$) to 2.88 ($SD = 0.99$). Mathematical computation saw similar growth for this group of students with a mean score of 2.58 ($SD = 1.06$) at pretest and 2.70 ($SD = 1.23$) at posttest. Students found most difficult the mathematical reasoning domain (pretest $M = 2.13$, $SD = 1.12$; posttest $M = 2.45$, $SD = 1.28$).

It is not surprising that students performed well in mathematical vocabulary, as teachers often find this domain easiest to implement in their instruction (Bravo et al., 2014). Students learned new vocabulary through direct instruction of new vocabulary but also strategies to make sense of unfamiliar mathematical terms, such as breaking the word apart into its morphological parts and using their knowledge of Spanish to make sense of the word.

While students wrote their responses in English, there was evidence of translanguaging as students wrote their responses. Some students wrote their entire response in Spanish, even though the prompt was in English. Other students put quotation marks around words they used in Spanish. Utilizing their bilingualism allowed students to show their full understanding of the mathematical concepts (see Gregory, this volume). Similarly, prompting students to use space to show their work utilizing mathematical visual literacy (e.g., labeled number line, graph) provided an additional scaffold that allowed students to show their full potential.

Mathematical reasoning was the most challenging element of the assessment for the 4th/5th grade EBLs in the study. While they did include mathematical reasoning in their responses, it was often only partially explaining or justifying their solution with limited examples of evidence to support their solution. In Figure 2.1 we offer an example of the challenges EBLs experienced with mathematical reasoning.

We see for this student an attempt to solve this mathematical problem concerning area but the student leverages his knowledge of base 10 to attempt to solve the problem. While not the most efficient model to utilize, the student does get partial credit for attempting to provide evidence to his reasoning regarding the approach he took to solve the problem.

4TH GRADE MATH WRITING PROMPT

Directions: Use the box to show your math work. In the lines below, explain your answer.

Your classroom needs a new carpet. The principal has asked you to find out how much carpet he will need to buy. The size of your classroom is 35 feet by 52 feet. Write an equation to represent the problem and then solve the problem.

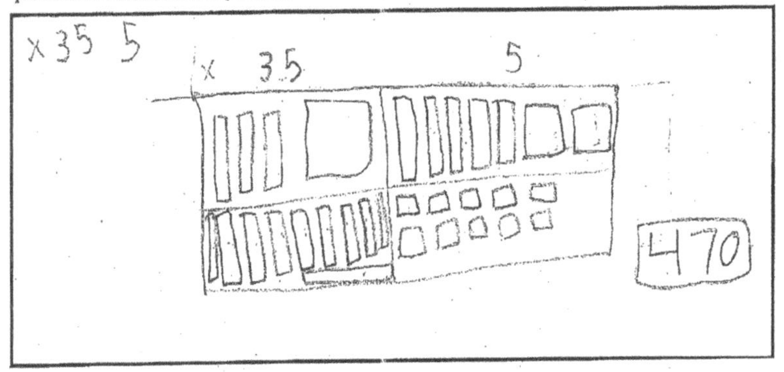

Explain to the principal how you found your answer.

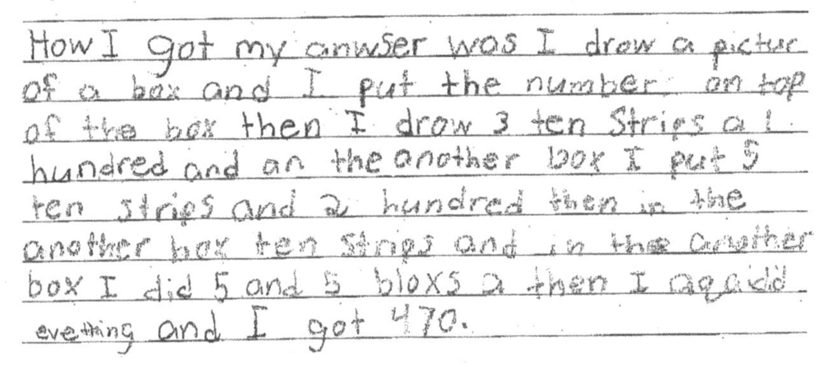
How I got my anwser was I draw a pictur of a box and I put the number on top of the box then I draw 3 ten Strips a 1 hundred and an the another box I put 5 ten strips and 2 hundred then in the another box ten strips and in the another box I did 5 and 5 bloxs 2 then I added eveything and I got 470.

Figure 2.1 4th grade writing sample.

Kindergarten/2nd Grade EBLs

While the fourth and fifth grade classrooms opted to administer the writing assessment in English, the Kindergarten and 2nd grade classrooms administered the writing assessment in Spanish. At the pretest, EBLs in Kindergarten and 2nd grade were weakest in mathematical vocabulary (pretest $M = 1.80$, SD 1.80; posttest $M = 2.03$, $SD = 1.07$) and strongest in mathematical literacy (pretest $M = 3.19$, $SD = 0.70$; posttest $M = 3.03$, $SD = 0.65$). EBLs made the strongest gains in mathematical computation (pretest 2.58, $SD = 1.06$; posttest 2.70, $SD = 1.23$).

Kindergarten and second grade EBLs used correct mathematical procedures to solve the mathematical problem. They used addition and subtraction appropriately and accurately. Yet, students did struggle utilizing mathematical vocabulary in their responses. This is likely due to kindergarten students still developing their writing skills and did not write the vocabulary words they knew during the writing assessment. In Figure 2.2 we offer one example from a student's response.

In Kimberly's response, we see strong visual literacy skills as she illustrates the number of balls for each student. Yet, in her response she writes *"Yo hice pelotas"* ("I made balls") with no mathematical vocabulary that might be expected (e.g., *sumar*/add; *total*/total).

Students utilized all of their linguistic resources to make meaning with print as they encountered the mathematics problems. It is clear that across the grade levels, more instructional support is needed in the area of mathematical reasoning. Instruction may include modeling of how to construct

KINDER: ESCRITURA MATEMATICA

Direcciones: Usa el espacio de abajo para mostrar tu trabajo de matemáticas.
En las líneas de abajo, explica tu respuesta.

Carlos tiene 3 pelotas y Ángel tiene 2 pelotas. ¿Cuántas pelotas hay en total?

Haz un dibujo que te ayude a resolver el problema.

Figure 2.2 4th grade writing sample.

a logical response with evidence to support their mathematical problem-solving strategy (Celedón-Pattichis & Turner, 2012). Yet, there is promise for utilizing writing to support mathematical learning and using writing to gauge what students know about the mathematics problems they are being asked to solve.

LIMITATIONS

Interpretations of these mathematical writing responses must consider several limitations. First, we did not have a comparison group to help control for factors that may have influenced the study results. Opportunities to conduct classroom observations to capture the quality and quantity of writing instruction EBLs were receiving during mathematics instruction. Finally, capturing the kindergarten's oral responses as to how they would solve the mathematical problem would have helped us gauge the level of vocabulary knowledge students possessed but could not write in their responses.

CONCLUSION

The use of this assessment to improve mathematics instruction in the language(s) of instruction fostered a responsive learning community where students' literacy paths led to enduring understandings about writing and mathematics (Garcia, 1999). The results from the corpus of data collected for the cohort of EBLs during their enrollment in their DLP school revealed three clear trends. First, when given the option to write in either language, students have an opportunity to demonstrate the depth of their mathematical knowledge (Moschkovich, 2013). EBLs' written products in turn provided a window for their teacher to see how students are understanding mathematics. Second, students had significantly more difficulty dealing with the demands of the mathematical reasoning and mathematical literacy domains than the mathematical computation and mathematical vocabulary domains. This clearly illustrates the need for more explicit and systematic instruction on representing mathematical understandings (literacy) and how to leverage evidence (reasoning) to support mathematical explanations. Lastly, there is great potential for integrating writing and mathematics (Bicer et al., 2012). We would argue it is critical to do so given that EBLs are tasked with not only learning content like mathematics but to also build their bilingual proficiency. The synergistic effect of integrating language and mathematical content allows EBLs to sharpen both their language and mathematics proficiency.

REFERENCES

Aguirre, J. M., Mayfield-Ingram, K., & Martin, D. B. (2013). *The impact of identity in K–8 mathematics: Rethinking equity-based practices.* National Council of Teachers of Mathematics.

Ashlock, R. B. (2006). *Error patterns in computation: Using error patterns to improve instruction.* Prentice Hall.

Basterra, M. R. (1998). *Excellence and equity for language minority students: Critical issues and promising practices.* The Mid-Atlantic Equity Consortium.

Bauer, B. E., & Gort, M. (2012). *Early biliteracy development: Exploring young learners' use of their linguistic resources.* Routledge.

Bicer, A., Capraro, R. M., & Capraro, M. M. (2013). Integrating writing into mathematics classrooms to increase students' problem-solving skills. *International Online Journal of Educational Sciences, 5*(2), 361–369.

Borasi, R., & Rose, B. J. (1989). Journal writing and mathematics instruction. *Educational Studies in Mathematics, 20,* 347–365.

Bravo, M. A., Mosqueda, E., Solís, J. L., & Stoddart, T. (2014). Possibilities and limits of integrating science and diversity education in preservice elementary teacher preparation. *Journal of Science Teacher Education, 24,* 22–51.

Casa, T. M., Firmender, J. M., Cahill, J., Cardetti, F., Choppin, J. M., Cohen, J., Cole, S., Colonnese, M. W., Copley, J., DiCicco, M., Dieckmann, J., Dorl, J., Garvin, M. K., Hebert, M. A., Karp, K. S., LaBella, E., Moschkovich, J. N., Moylan, K., Olinghouse, N. G., . . . Zawodniak, R. (2016). *Types of and purposes for elementary mathematical writing: Task force recommendations.* http://mathwriting.education.uconn.edu

Celedón-Pattichis, S., & Turner, E. (2012). "Explícame tu respuesta": Supporting mathematical discourse in emergent bilingual kindergarten students. *Bilingual Research Journal, 35*(2), 197–216.

Chval, K. B., Smith, E., Trigos-Carrillo, L., & Pinnow, R. J. (2021). *Teaching math to multilingual students: Positioning English learners for success.* Corwin.

Cummins, J. (2000). *Language, power, and pedagogy: Bilingual children in the crossfire.* Multilingual Matters.

de Araujo, Z., & Smith, E. (2022). Examining English language learners' learning needs through the lens of algebra curriculum materials. *Educational Studies in Mathematics, 109,* 65–87.

Dyson, A. H. (1993). *Social worlds of children learning to write in an urban primary school.* Teachers College Press.

Garcia, E. E. (1999). *Understanding and meeting the needs of culturally diverse students.* Houghton Mifflin Co.

Meel, D. (1999). Email dialogue journals in a college calculus classroom: A look at the implementation and benefits. *Journal of Computers in Mathematics and Science Teaching, 18*(4), 387–413.

Moll, L. C. (2001). The diversity of schooling: A cultural-historical approach. In M. de la Luz Reyes, & J. J. Halcon (Eds), *The best for our children: Critical perspectives on literacy for Latino students.* Teachers College Press.

Morgan, C. (1994). Writing mathematically. *Mathematics Teaching, 146,* 18–21.

Moschkovich, J. N. (2013). Principles and guidelines for equitable mathematics teaching practices and materials for English language learners. *Journal of Urban Mathematics Education, 6*, 45–57.

Musanti, S. I., Celedón-Pattichis, S., & Marshall, M. E. (2009). Reflections on language and mathematics problem solving: A case study of a bilingual first-grade teacher. *Bilingual Research Journal, 32*, 25–41.

National Center for Education Statistics. (2023). English learners in public schools. *Condition of Education.* U.S. Department of Education, Institute of Education Sciences. Retrieved on January 21, 2024 from https://nces.ed.gov/programs/coe/indicator/cgf

National Governors Association Center for Best Practices & Council of Chief State School Officers. (2010). *Common Core State Standards for Mathematics.*

Pugalee, D. K. (2004). A comparison of verbal and written descriptions of students' problem- solving processes. *Educational Studies in Mathematics, 55*(1), 27–47.

Rubinstein-Ávila, E., Sox, A. A., Kaplan, S., & McGraw, R. (2015). Does biliteracy+ mathematical discourse = binumerate development? Language use in a middle school dual-language mathematics classroom. *Urban Education, 50*(8), 899–937.

Sierpinska, A. (1998). Three epistemologies, three views of classroom communication: Constructivism, sociocultural approaches, interactionism. In H. Steinberg, M. G. B. Bussi, & A. Sierpinska (Eds.), *Language and communication in the mathematics classroom* (pp. 30–62). National Council of Teachers of Mathematics.

Valdés, G. (2005). Bilingualism, heritage language learners, and SLA research: Opportunities lost or seized? *The Modern Language Journal, 89*, 410–426.

Paris, D. (2012). Culturally sustaining pedagogy: A needed change in stance, terminology, and practice. *Educational Researcher, 41*(3). https://doi.org/10.3102/0013189X12441244

U.S. Government Accountability Office. (2022). *Back to school for K–12 students: Issues ahead.* Report to Congressional Requesters.

CHAPTER 3

BIDS FOR LINGUISTIC CAPITAL THROUGH TRANSLANGUAGING DURING SPANISH AND ENGLISH MATHEMATICS INSTRUCTION IN DUAL LANGUAGE CLASSROOMS

Jolene Castillo Gregory
Denver Public Schools

ABSTRACT

This chapter explores how student translanguaging rates vary as a function of teachers' response to translanguaging, school context, and student language proficiency within mathematics instruction in elementary dual language programs (DLP). The data are drawn from the translanguaging practices of students and teachers in 30 elementary Spanish-English DLP classrooms in Texas and California. Statistically significant variations in translanguaging rate were

Mathematics Instruction in Dual Language Classrooms, pages 41–62

www.infoagepub.com

found as a function of location, teacher response, language of instruction, student language proficiency and grouping of the students. The findings challenge previous research regarding teachers' responses to translanguaging in DLP and the relationship between language proficiency and translanguaging, in addition to underscoring the capital and symbolic power that the language of power bestows. The findings suggest that DLP educators' approach translanguaging from a new perspective than previously reported by research.

In light of the increasing numbers of emerging bilingual children in U.S. schools (U.S. Department of Education, 2019), research on teacher and student language use, and the effects of language use on students' learning has become more urgent. One practice being promoted to support student learning is translanguaging (Garcia & Wei, 2015; Moschkovich, 2007; Palmer et al., 2014; Planas, 2018). Translanguaging generally refers to instances when bilingual speakers use their two languages in some combination to communicate. While previous researchers and educators may have used the term "code-switching" when describing this phenomenon, the term code-switching tends to consider distinct languages as separate systems while translanguaging views a bilingual speakers' full linguistic repertoire as an integrated system (Canagarajah, 2011). Some advocates of translanguaging consider distinct "languages" as merely socio-political constructions (Makoni & Pennycook, 2007). Nevertheless, translanguaging (and code switching) appears to serve social (Giles & Ogay, 2007; Lin, 2013; Martínez, 2009; Myers-Scotton, 1995; Sommerville & Faltis, 2019), linguistic (Cenoz, 2017; Creese & Blackledge, 2010; Gort & Sembiante, 2015; Lin, 2013), and academic (García, 2013; Gort & Pontier, 2013; Palmer et al., 2014) functions, but how these functions interact is largely unknown.

The disagreements regarding the differences between translanguaging and code-switching have been well documented (see MacSwan, 2017, for a review) and my purpose here is not to resolve those differences, but rather to better understand how two languages are being used in mathematics DLP classrooms and the factors influencing their use (see Lindholm-Leary, 2013 for a review).

I have chosen to study translanguaging in DLP programs primarily because the students in such programs are learning two languages by design, and all students are bilingual to some degree. Research and the promotion of translanguaging has taken place in contexts where not all participants are bi/multilingual in the same languages, which leads to certain pragmatic opposition (McCarthy, 2018) because not all interlocutors can engage with one another through translanguaging. One option to reduce the pragmatic opposition is to conduct research in DLP. Another advantage is that DLP programs have adopted various policies regarding translanguaging, making them a compelling context for study. Some DLP programs have maintained a strict separation of the two languages, announcing to the students, for instance, "It's

Spanish time now, so no speaking English" or "Goodbye Spanish, hello English!" Indeed, the formal policies regarding translanguaging in DLP seem to represent one of three positions. One holds that translanguaging should be *prohibited* (e.g., McCarthy, 2018; Wang & Kirkpatrick, 2013) as reflected in the example just mentioned; another that it should be *permitted* (e.g., Macaro, 2005; Setati et al., 2002; Weber, 2014); and third, that it should be *promoted* (e.g., García & Kleyn, 2016; Sommerville & Faltis, 2019). The latter position argues, among other points, that it is unfair to ask students to circumscribe their full linguistic repertoire by limiting the use of both languages when students may find it useful (García & Kleyn, 2016; Otheguy et al., 2015). However, elementary school students often ignore or circumvent school policies on a variety of issues; therefore, the actual use of translanguaging has been shown to be much less straightforward and that other factors (e.g., language status) influence language use (Planas & Setati, 2009). Previous research has found that some teachers prohibit translanguaging (e.g., Sommerville & Faltis, 2019), but it is unclear if students were permitted to translanguage, with what frequency, or in what contexts they might use translanguaging, and which of the functions it serves.

This study documents elementary school students' use of two languages (Spanish and English) during mathematics instruction in 30 different DLP classrooms across two states (California and Texas) and several school districts, as well as interviewing 14 of the teachers in whose classrooms the data were collected. By documenting the language use of teachers and students across a range of settings, this approach will provide the reader an opportunity to consider language use and the implications on mathematics instruction. The teachers' interpretations of how and why students make use of their languages offers implications for DLP programming and informs the debate on whether translanguaging is promoted, permitted, or prohibited in DLP.

Most studies of translanguaging in DLP classrooms (García & Sylvan, 2011; Gort & Pontier, 2013; Palmer et al., 2014) have been conducted during language arts instruction (King & Ridley, 2019; Martínez et al., 2015). The exploration of translanguaging in mathematics instruction provides an opportunity to explore translanguaging when instruction takes place in different languages in the same context. In addition, mathematics instruction in DLP is a linguistically and pedagogically complex environment where dilemmas surround the use of translanguaging and other language practices in the classroom. Teachers face dilemmas such as how to develop the language and the subject matter content, whether to foreground the language or the subject matter content, as well as considering the social and political implications surrounding students' and teachers' language choices and opportunities that exist in the classroom (Adler, 1998).

Furthermore, the study of translanguaging cannot be separated from social and cultural dimensions and exploring translanguaging in mathematics classrooms (Civil, 2010) may contribute to new insights into translanguaging, content learning, and the social implications of language use. Finally, by studying a wide range of classrooms and teachers (which is generally uncommon in the translanguaging research), I hope the study will offer more generalizable findings.

This study is guided by four questions:

1. How does the translanguaging rate vary as a function of teachers' responses to translanguaging in Spanish-English DLP?
2. How does translanguaging rate vary as a function of school context in Spanish-English DLP?
3. How does translanguaging rate vary as a function of student Spanish and English language proficiency?
4. When teachers are shown the data showing their students' language use, what factors do they believe account for their translanguaging?

The following sections explore the research on mathematics instruction in DLP and translanguaging in mathematics teaching and learning, followed by the study's methods, results, and conclusions.

LANGUAGE AND MATHEMATICS LEARNING IN DLP

DLPs strive to develop students' linguistic proficiency in two languages simultaneously with academic performance at or above grade level, as well as develop positive cross-cultural behaviors and attitudes (Howard et al., 2005). Because mathematics and language learning are inseparable and interrelated (Brown, 2002), DLP programs must focus on developing both simultaneously. As such, mathematics instruction in DL programs is grounded in theories from the fields of education (Martinez et al., 2015) and applied linguistics. The language differentiation model (Volterra & Taescher, 1978) particularly influences language policy, instruction, and translanguaging. Drawing on the language differentiation model (Volterra & Taescher, 1978), DL teachers promote linguistic purism and language separation because they believe the students' use of one language interferes with and causes a disuse of the other language, and therefore impedes the development of the other language (Lindholm-Leary, 2006; Martínez et al., 2015; Nambisan, 2014; Torres-Guzmán, 2007). DL teachers also emphasize the standard language variety (Sommerville & Faltis, 2019) as it is seen as the proper language necessary for academic performance, particularly in mathematics. Mastery of the specialized words, symbols, math register (Halliday & Martin, 1993; Pimm, 1987); discourse (Gee, 1996); and modes of argument such as precision, brevity, and logical coherence (Forman, 1996) are considered necessary in mathematics so as to be able to participate in mathematics discourse communities (Roth & Tobin, 2007; Solomon, 2009). Mastery of these elements also indicate mathematical competence and school success (Celedón-Pattichis et al., 2010; Lemke, 1990; Martínez et al., 2015) and thus translanguaging may be discouraged. These theories have led to language separation or monolingual language (in the language of instruction) policies in DL mathematics classrooms.

However, strict language separation policies have also been critiqued based on empirical evidence demonstrating the value of using the first language for second language acquisition (Moore, 2013) and that asking bilingual children to perform with less than half their repertoire is unfair and not supported by research on second language development (García & Kleyn, 2016). In addition, translanguaging has been found to aid in literacy skill development (Martínez-Álvarez, 2017) and content learning (Alvarez, 2012). Some researchers also promote an unrestricted use of translanguaging in the classroom from a transformative stance. They argue for using a child's full linguistic repertoire to transform language hierarchies in schools (García & Kleyn, 2016). Sommerville and Faltis (2019) likewise argue that translanguaging in schools that promote language separation can be understood as a tactic (de Certeau, 1984) that challenges the traditional language separation policy.

Studies promoting the use of translanguaging suggest that the direction of translanguaging would be from the language of instruction to the home language because it is associated with students being more relaxed and making connections to the mathematical procedures, concepts, and vocabulary they are learning, which enhances their understanding (Bose & Choudhury, 2010; Planas & Civil, 2013; Salehmohamed & Rowland, 2014; Setati, 2005; Setati et al., 2002; Tavares, 2015). However, research in contexts outside of the United States has revealed that this is not always the case (e.g., Setati, 2008; Planas & Setati, 2009).

METHODOLOGY

Setting and Participants

A naturalistic inquiry (Patton, 2002) approach was taken in order to explore the use of student and teacher translanguaging practices. The data were collected from 30 classrooms (2 classrooms participated both years) during the 2018–2019 and 2019–2020 school years (pre COVID-19 remote teaching) in nine different urban school districts in California and Texas. All of the schools taught the DLP using a "90/10" model in the early grades, moving to a 50/50 model in later grades (typically fourth grade). Each classroom enrolled between 20–30 students, with varying ratios of English dominant, Spanish dominant and simultaneous bilingual students. The teachers were recruited and participated in a funded professional development project focusing on vocabulary, discourse and literacy development in DLP mathematics classrooms,[1] but translanguaging was not an explicit focus of the project learning. All teachers were certified bilingual teachers (advanced language skills in Spanish and English). Fourteen teachers from these classrooms were chosen at random to be interviewed regarding their beliefs about the language use documented in their classrooms.

Table 3.1 describes the grades, state location, language of instruction for mathematics, teacher years of experience, and school demographics.

DATA COLLECTION AND CODING

The dataset was compiled through lesson videos and semi-structured interviews with the teachers in 14 of the classrooms chosen at random (see classrooms marked with an asterisk in Table 3.1). The interviews explored the teachers' beliefs about the language use documented in their classrooms. The videos ranged from 10 to 60 minutes per teacher (lessons in the lower grades were shorter), for a total of 20 hours and 31 minutes of classroom videos. Each teacher wore a microphone to record their voices and the voices of the children near them. Following the approach taken by Brevik and Rindal (2020), I focused on only those translanguaging instances recorded by the teacher's microphone. Although I recognize that not all translanguaging instances in the classroom were detected by the microphone, a choice was made by the teacher about how to respond to the translanguaging that was picked up by the microphone (i.e., heard by the teacher). Because teacher response was a variable in the model, only those within hearing distance of the teacher were considered.

The coding of the video recordings aimed to identify the translanguaging instances, teacher ID, state, time stamp of the instance, duration of video, duration of translanguaging instance, language of instruction, translanguager, student grouping, teacher response to translanguaging, grade level, state, and a simple transcription of the translanguaging instance. I drew on Myers-Scotton's (2006) matrix language framework to identify each instance. In this framework, Myers-Scotton (2006) identifies an embedded language (or guest language), which is inserted in the matrix language. Although Myers-Scotton (2006) uses grammar to determine the matrix language, I took a sociolinguistic approach and used the context to determine the matrix language. I considered the target language during the mathematics class as the matrix language and the translanguaging instance was described as beginning with the use of an embedded language and ending

TABLE 3.1 Participating Classrooms

Classroom	Grade	State	Language of Instruction for Mathematics	Teacher's Years of Experience	School Demographics[a]
1*	K	CA	Spanish Only	5	10.5% Asian, 0.3% Black, 14.7% Filipino, 74% Hispanic, 1% Native Hawaiian, 2.6% Multiple, 2.6% White, 63% FRL, 54% MLL
2*	K	CA	English Only	4	23% Asian, 1% Black, 46% Hispanic, 7% Multiple, <1% Native Hawaiian, <1% Native American, 23% White, FRL and MLL Not available
3	K	CA	English Only	20	23% Asian, 1% Black, 46% Hispanic, 7% Multiple, <1% Native Hawaiian, <1% Native American, 23% White, FRL and MLL Not available
4	K	TX	Spanish Only	18	0.2% Asian, 2.1% Black, 96.8% Hispanic, 0.9% White, 88% FRL, 67% MLL
5*	1	CA	Spanish Only	1	0.5% Asian, 0.5% Black, 0.2% Filipino, 95.8% Hispanic, 0.2% Multiple, 1.9% White, 71% FRL, 7% MLL
6*	1	CA	Spanish Only	3	0.5% Asian, 0.5% Black, 0.2% Filipino, 95.8% Hispanic, 0.2% Multiple, 1.9% White, 71% FRL, 7% MLL
7*	1	CA	English Only	5	10.5% Asian, 0.3% Black, 14.7% Filipino, 74% Hispanic, 1% Native Hawaiian, 2.6% Multiple, 2.6% White, 63% FRL, 54% MLL
8	1	CA	Spanish Only	18	0.5% Asian, 0.5% Black, 0.2% Filipino, 95.8% Hispanic, 0.2% Multiple, 1.9% White, 71% FRL, 7% MLL
9	1	TX	Spanish Only	26	1% Asian, 1% Black, 91% Hispanic, 1% Multiple, <1% Native American, <1% Native Hawaiian, 6% White, 98% FRL, 8% MLL
10*	2	CA	Spanish Only	17	0.6% Asian, 0.6% Black, 0.4% Filipino, 95.6% Hispanic, 1.1% Multiple, 1.7% White, 72% FRL, 56% MLL
11	2	TX	Spanish Only	10	<1% Asian, 5% Black, 67% Hispanic, 2% Multiple, <1% Native American, <1% Native Hawaiian, 26% White, 72%, FRL, 40% MLL

(continued)

TABLE 3.1 Participating Classrooms (continued)

Classroom	Grade	State	Language of Instruction for Mathematics	Teacher's Years of Experience	School Demographicsa
12*	3	CA	Primarily English w/ some Spanish	1	0.5% Asian, 0.5% Black, 0.2% Filipino, 95.8% Hispanic, 0.2% Multiple, 1.9% White, 71% FRL, 7% MLL
13*	3	TX	Primarily Spanish w/ some English	14	0.2% Asian, 2.1% Black, 96.8% Hispanic, 0.9% White, 88% FRL, 67% MLL
14	3	TX	English Only	7	0.2% Asian, 1.5% Black, 91.1% Hispanic, 1.2% Multiple, 0.2% Native Hawaiian, 5.8% White, 58% FRL, 47% MLL
15*	3	TX	Primarily Spanish w/ some English	10	7% Asian, 3% Black, 67% Hispanic, 3% Multiple, 20% White, 36% FRL, 25% MLL
16*	3	TX	Spanish Only	15	0.2% Asian, 1.5% Black, 91.1% Hispanic, 1.2% Multiple, 0.2% Native Hawaiian, 5.8% White, 58% FRL, 47% MLL
17*	3	TX	Spanish Only	15	0.2% Asian, 2.1% Black, 96.8% Hispanic, 0.9% White, 88% FRL, 67% MLL
18	4	CA	Primarily Spanish w/ some English	10	0.6% Asian, 0.6% Black, 0.4% Filipino, 95.6% Hispanic, 1.1% Multiple, 1.7% White, 72% FRL, 56% MLL
19	4	TX	Spanish Only	12	<1% Asian, 5% Black, 67% Hispanic, 2% Multiple, <1% Native American, <1% Native Hawaiian, 26% White, 72% FRL, 40% MLL
20*	4	TX	Primarily English w/ some Spanish	7	<1% Asian, 5% Black, 67% Hispanic, 2% Multiple, <1% Native American, <1% Native Hawaiian, 26% White, 72% FRL, 40% MLL
21	4	CA	English Only	4	0.5% Asian, 0.5% Black, 0.2% Filipino, 95.8% Hispanic, 0.2% Multiple, 1.9% White, 71% FRL, 7% MLL
22	4	CA	English Only	15	0.2% Asian, 1.3% Black, 0.6% Filipino, 89% Hispanic, 1.7% Multiple, 6.4% White, 70% FRL, 52% MLL

(continued)

TABLE 3.1 Participating Classrooms (continued)

Classroom	Grade	State	Language of Instruction for Mathematics	Teacher's Years of Experience	School Demographics[a]
23*	4	TX	Primarily English w/ some Spanish	25	1% Asian, 1% Black, 91% Hispanic, 1% Multiple, <1% Native American, <1% Native Hawaiian, 6% White, 98% FRL, 8% MLL
24	4	TX	English Only	8	1% Asian, 1% Black, 91% Hispanic, 1% Multiple, <1% Native American, <1% Native Hawaiian, 6% White, 98% FRL, 8% MLL
25*	5	CA	Spanish Only	23	0.1% American Indian, 3.8% Asian, 2.1% Black, 0.1% Filipino, 49.7% Hispanic, 0.4% Native Hawaiian, 9.3% Multiple, 34.2% White, 28% FRL, 18% MLL
26	5	CA	English Only	1	0.5% Asian, 0.5% Black, 0.2% Filipino, 95.8% Hispanic, 0.2% Multiple, 1.9% White, 71% FRL, 7% MLL
27	5	CA	English Only	5	0.8% American Indian, 1.1% Black, 97% Hispanic, 0.8% Native Hawaiian, 0.4% White, 85% FRL, 53% MLL
28	5	CA	English Only	5	9.4% Asian, 1.6% Black, 0.8% Filipino, 79% Hispanic, 1.9% Native Hawaiian, 3.5% Multiple, 2.9% White, 65% FRL, 47% MLL
29	5	CA	Spanish Only	1	0.8% American Indian, 1.1% Black, 97% Hispanic, 0.8% Native Hawaiian, 0.4% White, 85% FRL, 53% MLL
30	6	CA	Primarily English w/ some Spanish	3	100% Hispanic, 93% FRL, 75% MLL

Note: FRL = Free and Reduced-Price Lunch, MLL = Multiligual Learner

* indicates classrooms where the teacher was interviewed. Teacher 28 was the only one who identified as male. All others identified as female.

[a] The classrooms in California were located in the San Francisco bay area, while the Texas classrooms were located near San Antonio.

when the speaker returned to the matrix language. In some circumstances the speaker switched back and forth between both the matrix language and the embedded language. In this case, an instance began with the first use of the embedded language and ended when they switched back to the matrix language for a complete sentence. Codes from the semi-structured interviews with the teacher were then added to each coded translanguaging instance (see Appendix for a complete description of each code). The dataset included 447 translanguaging instances after extreme outliers (RATE > 0.254) were removed.

After coding the videos, I conducted semi-structured interviews (Kvale, 1996) with 14 of the 32 teachers. Random selection of the teachers yielded a range of translanguaging rates, states, school districts, and language of instruction. This allows for exploration of the previously described relationship between translanguaging and the context (Borg, 2003; Brown & Cooney, 1982; Nava, 2009; Skott, 2015), social interactions (Flores, 2001), and the political affordances and constraints of teachers' situations (Windshitl, 2002). The semi-structured interviews using video stimulated recall (Patthoff, 2021) explored the instructional setting including language of instruction, state, grouping of students, mathematics problem type, school and teacher language policy, and student language proficiency for the translanguaging instances identified in the video recordings.

VARIABLES IN THE MODELS

Question Predictors

The language of instruction (LOI) is a dichotomous nominal variable coded 0 = English, 1 = Spanish. The mean for LOI is 0.82 ($sd = 0.37$); that is; 82% of the lessons recorded were taught in Spanish. The student's English language proficiency (SELP) and Spanish language proficiency (SSLP) are ordinal variables that range from 1 to 3 (low to high), based on the teacher-reported level of student language proficiency. The mean for SELP is 2.47 ($sd = 0.89$). The mean for SSLP is 2.42 ($sd = 0.77$).

Outcome Variable

Teachers' response (TRESP) is an ordinal variable coded 0 = prohibit, 1 = permit, 2 = promote. The mean of TRESP is 1.22 ($sd = 0.48$). Translanguaging rate (RATE) was calculated by dividing the duration of the translanguaging instance by the length of the lesson video in seconds. The mean for RATE is 0.08 ($sd = 0.06$).

Control Predictors

The analyses included control predictors to account for the political and classroom context variables (see Table 3.2). State (STATE) is a dichotomous nominal variable coded 0 = California, 1 = Texas. The mean for STATE is 0.64 (sd = 0.48). Student grouping (GROUP) is a dichotomous nominal variable coded 1 = whole class, 2 = small group. The mean for GROUP is 1.44 (sd = 0.5).

DATA ANALYSIS

Analysis identified extreme outliers with a translanguaging rate > 0.254 (2 *sd*), which were removed from the dataset. Data analysis used ordinary least squares (OLS) regression to examine the main effects of how the translanguaging rate varies as a function of teachers' response to translanguaging, how the translanguaging rate varies as a function of school context, and how the translanguaging rate varies as a function of students' language proficiency in English and Spanish when the language of instruction was Spanish. It was not possible to examine the relationship between translanguaging rate and students' language proficiency when the language of instruction was English (n = 17) or the relationship between translanguaging rate and teachers' language proficiency in English (n = 40) and Spanish

TABLE 3.2 Descriptive Statistics of all Variables in Models

Variable	Description	*N*	mean	*sd*	min.	max.
RATE	Translanguaging rate (translanguaging duration in seconds/length of video in seconds)	447	0.08	0.06	0	0.25
TRESP	Teacher response to translanguaging 0 = prohibit, 1 = permit, 2 = promote	445	1.22	0.48	0	2.00
LOI	Language of instruction 0 = English, 1 = Spanish	447	0.82	0.39	0	1.00
GROUP	Student grouping 0 = whole class, 1 = small group	440	0.44	0.5	1	2.00
STATE	State 0 = California, 1 = Texas	447	0.64	0.48	0	1.00
SSLP	Teacher Rating of Student Spanish Language Proficiency 1 = low, 2 = average, 3 = high	104	2.42	0.77	1	3.00
SELP	Teacher Rating of Student English Language Proficiency 1 = low, 2 = average, 3 = high	101	2.47	0.9	1	3.00

($n = 42$) due to the limited number of instances in the sample. An OLS regression was run with the predictor variables and correlations examined between the categorical variables.

The linear regression model was:

$$\text{RATE} = \beta_0 + \beta_1\text{LOI} + \beta_2\text{GRADE} + \beta_3\text{STATE} + \beta_4\text{GROUP} + \beta_5\text{TRESPPROMOTE} + \beta_6\text{TRESPPROHIBIT} + \beta_7\text{SSLP} + \beta_8\text{SELP} + \varepsilon$$

FINDINGS

Model 1 and 2 serve as the control models of language of instruction and student English proficiency without covariates, and Models 3 and 4 include covariates of the school context and the student English proficiency (see Table 3.3). In examining the overall explanatory value of each model using R^2, it is evident that the inclusion of covariates increases the explanatory value in Models 3 and 4. Model 4 has the greatest explanatory value ($R^2 = .80$) and examines the main effects of both the school context and the student English proficiency on translanguaging rate.

TABLE 3.3 OLS Regression Models Predicting Translanguaging Rate

	Mode 1	Model 2	Model 3	Model 4
Fixed Effects	Coef. (SE)	Coef. (SE)	Coef. (SE)	Coef. (SE)
Intercept	.018* (.005)	.095*** (.009)	.085*** (.009)	.037** (.012)
School Context				
LOI–Spanish	.078 (.006)		.014* (.006)	.029*** (.007)
Grade			–.010*** (.001)	–.010*** (.001)
Texas			.059*** (.004)	.051*** (.004)
Small Group			–.014*** (.003)	.009* (.004)
TRESP–Promote			–.040*** (.005)	–.042*** (.009)
TRESP–Prohibit			.005 (.010)	–.029 (.016)
Student Language Proficiency				
English		–.009* (.004)		–.003 (.002)
Regression Values				
SE	0.047	0.032	0.035	0.015
R^2	0.29	0.07	0.62	0.80
F	177.74	6.89	118.10	53.15
p	0.000	0.010	0.000	0.000

* $p < .05$; ** $p < .01$; *** $p < .001$

The parameter estimates indicate that the translanguaging rate on average was .48 standard deviations higher ($\beta = .029$, 95% CI [.015, .062], $p < .000$) when the language of instruction was Spanish than when it was English. Secondly, they indicate that the translanguaging rate on average was .16 standard deviations lower ($\beta = -.010$, 95% CI [–.012, –.008], $p < .000$) for each successive grade higher. Third, they indicate that the translanguaging rate on average was .85 standard deviations higher ($\beta = .051$, 95% CI [.043, .059], $p < .000$) in Texas than in California. In addition, they indicate that the translanguaging rate was .15 standard deviations higher ($\beta = .009$, 95% CI [.002, .017], $p < .000$) when the students were in small groups than when they were in whole class. Fifth, they indicate that the translanguaging rate on average was .7 standard deviations lower ($\beta = -.042$, 95% CI [–.060, –.002], $p < .05$) when the teacher response was promoting translanguaging when compared to permitting and prohibiting translanguaging. Strong correlations were observed between teacher response of promoting and language of instruction of English ($R = .579$, $p < .01$). Finally, they indicate that the translanguaging rate on average was .05 standard deviations lower ($\beta = -.003$, 95% CI [–.007, .000], $p < .089$) for each unit increase in student English proficiency. Student Spanish proficiency was not found to be a statistically significant predictor of translanguaging rate. The final model of how translanguaging rate varies as a function of school context and student English proficiency is:

$$\text{RATE} = 0.37 + 0.029(\text{LOI}) - 0.01(\text{GRADE}) + 0.51(\text{STATE}) + 0.009(\text{GROUP}) - 0.042(\text{TRESPPROMOTE}) - 0.029(\text{TRESPPROHIBIT}) - 0.003(\text{SELP}) + \varepsilon$$

DISCUSSION/IMPLICATIONS

The following section addresses my findings as they are directly related to my research questions. The order in which I present the findings is based on the strength of the predictors in the model.

How does translanguaging rate vary as a function of school context in Spanish-English DLP?

The state where the classrooms were located (California or Texas) was the best predictor of translanguaging rate, with students in Texas having higher rates. This variation between the states, while perhaps interesting from a policy perspective, is a factor beyond the control of any educator interested in either promoting or prohibiting translanguaging (we cannot make kids move from one state to another). Nevertheless, a brief analysis suggests that perhaps the policy environment in California, where state Proposition 227 was passed, might discourage students from translanguaging. As Setati (2008) has suggested, language policies do influence students' language use. Another potential explanation is that schools in Texas allow students to take mandated tests in Spanish until the fifth grade, thus encouraging additional use of Spanish until students enter middle school in the sixth grade. Finally, the proximity of the schools to the Mexico–United States border, where translanguaging might be more common, could also explain why Texas students translanguaged more (Esquinca et al., 2014). It is also interesting to note that I found more translanguaging in whole class settings than in smaller groups, another finding that appears to run counter to previous research (Planas & Setati, 2009). A possible explanation for this is that students' performance in a whole class setting may garner them more power, as will be discussed a little later.

How does the translanguaging rate vary as a function of teachers' responses to translanguaging in Spanish-English DLP?

The second most influential predictor, and the more interesting one for DLP educators, was the teacher's stance towards translanguaging. First, I would point out that few teachers in the sample specifically prohibited translanguaging. In fact, some of the teachers permitted or encouraged translanguaging even if the school's DLP policies discouraged it. But the data revealed an even more surprising result: Students showed lower translanguaging rates when the teacher promoted translanguaging than when they permitted or prohibited it. This finding, which requires replication in other settings, appears to contradict previous work suggesting that teachers have great control over their student's translanguaging (García & Kleyn, 2016; García & Sylvan, 2011; Gort & Pontier, 2013; Otheguy et al., 2015; Palmer et al., 2014).

How does translanguaging rate vary as a function of student Spanish and English language proficiency?

I found that the language of instruction is a significant predictor of the translanguaging rate, with much higher rates when the language of instruction was Spanish, irrespective of students' English proficiency. That is, translanguaging rates (from Spanish to English) increase when the language of instruction is Spanish and even for students who have lower proficiency in English. By recognizing the "direction" of the translanguaging, I might be analyzing the findings within the code-switching traditions (i.e., proponents of translanguaging suggest the language direction is unimportant, as long as students are using their full linguistic repertoire). Regardless of the theoretical differences, the direction is important, as I will explain in the final question using this example.

The following classroom scenario is representative of this move from the teacher teaching in Spanish to students sometimes briefly speaking in Spanish but moving quickly to English, what I would call teacher-student translanguaging. It might be suggested that "teacher–student" translanguaging is not genuine translanguaging, but then what is it? Interpersonal translanguaging? In Classroom 25 (mathematics instruction in Spanish), Viviana (a pseudonym)—who the teacher indicated is a native Spanish speaker with high proficiency, but much lower proficiency in English—and Pia (also a pseudonym)—whose native language was English but who also had high proficiency in Spanish (again based on the teacher's assessment)—were working on the floor doing calculations on their white boards. The teacher reminded the students (in Spanish) to put commas in their numbers. Viviana turns to Pia and says, "one, two, three" and points to where Pia is missing commas. In this case, as in so many others, Viviana was choosing to translanguage from the language of instruction (Spanish) in which both she and Pia had high proficiency in order to help her classmate in her own less developed language. However, this general finding of moving to all English was not found in all instances. In the interaction below, we find the teacher teaching asking questions in Spanish and the student translanguaging:

T: *Que estas haciendo?* [What are you doing?]

Victoria: I have an *entero* [whole] and I put un *medio* [half] in the middle and then I put a fourth and then I put another fourth and it makes un entero

T: *Oh! Pusiste un medio y luego dos cuartos. Y que lograste?* [Oh! You put a half and then two quarters. And what did you get?

Victoria: Un entero. [a whole]

When teachers are shown the data showing their students' language use, what factors do they believe account for their translanguaging?

These complicated linguistic phenomena are found within elementary classrooms, where teachers often lead discussions and guide conversations, and, in the best of cases, orchestrate discourse that makes every student believe they made a contribution. In order to make sense of this type of translanguaging, I asked the teachers for their thoughts on why their students consistently went from hearing a mathematics lesson in Spanish and quickly moving to English to answer questions or work in small groups.

Overall, the teachers recognized the general pattern of language use in their mathematics lesson, but most were surprised by how often it occurred. In general, the teachers reported a social function: The proficient Spanish speakers wanted to help their peers by moving to English, as one suggested, "They [the more proficient Spanish speakers] are trying to help the English-speaking students" by speaking English, although each teacher struggled with a clear explanation ("I don't know, up. I'm trying to think, I'm trying to even think like maybe... like who they sit with?"). Pia and Viviana's teacher suggested yet another purpose:

> I don't know, maybe because she's [Viviana] an EL and she wants to prove that she can speak English... Some of the other kids, I think that that's what happens, that they're trying to prove, like, "Oh, I can speak in English just like everybody else." (Teacher, Classroom 25)

What is striking about these explanations is that they are rooted in the social function of translanguaging. As previous research has found (Bose & Choudhury, 2010; Planas & Civil, 2013, Salehmohamed & Rowland, 2014; Setati, 2005; Setati et al., 2002; Tavares, 2015), translanguaging is rarely "compensatory." Setati (2008), Planas (2018), and Setati (2009) point out that translanguaging can be used as a bid to gain capital or status through the use of the language of power. These bids often took place in whole-class settings as previously mentioned. Perhaps these very public displays of using English allotted the students more power than less public displays in small group settings. Considering the potential social functions of translanguaging, it is also perfectly logical that children would speak in a language that helps their friends or to gain the respect of their classmates through their use of language.

CONCLUSION

The findings from this study have important implications for practice, policy, and future research. But first, DLP educators may be better served trying to understand their students' translanguaging rather than trying to conform their translanguaging practices to theories or policies *about* translanguaging. Researching translanguaging in mathematics provided insights into how multilingual students are using their full linguistic repertoire in the classroom. Based on my data, the students (and often the teachers) are ignoring any school-wide regulations for prohibiting, permitting, or promoting translanguaging. Nearly all the teachers were permitting students to choose their language, which raises an important point. The students in this study were using English almost exclusively when speaking about mathematics and their purpose appears to be socially driven.

This study is not able to confirm why students are using language in this way, and therefore further research should explore the function of students' translanguaging in the mathematics classroom. Gort and Pontier (2013) suggest that educators "foster language practices that approximate authentic interactional contexts existing outside of school" (p. 240). To do this, we must come to understand the functions for which students use translanguaging and how it empowers students to engage with the society's linguistic "market." As research, theory, and practice support the social functions of translanguaging, students will benefit in and beyond the classroom. Further research is also necessary to deepen our understanding of the relationship between teachers' response, language proficiency, and language status with students' translanguaging practice, particularly during English instruction in DLP across content areas.

Despite the robust findings, several challenges to the study's validity can be raised. First, I used a measure of translanguaging that is more conceptually consistent with the work in code switching. Unfortunately, none of the research emerging from the translanguaging literature details a coding scheme that could yield a viable dependent variable for my study. Second, I did not talk to the students about their translanguaging. This will be a topic of my future research. Third, the context may limit the study's generalizability. If the same research were conducted in language arts, the results could differ. Fourth, the teacher-reported data in the study (e.g., student language proficiency) must be interpreted with caution. I recognize that language proficiency can be measured in different ways, and I had to trust the teachers to provide an accurate assessment. Fifth, I collected no achievement data to help answer whether translanguaging improves performance in mathematics.

APPENDIX

Video Coding Scheme

Primary Code	Description
Teacher ID (TID)	1–32
State (STATE)	0 = California, 1 = Texas
Time Stamp (TIME)	
Duration of instance (DURTL)	In seconds
Duration of Video (LENGTH)	In seconds
Translanguaging Rate (RATE)	Sum of the duration of translanguaging instances for the teacher/duration of video
Language of Instruction (LOI)	0 = English, 1 = Spanish
Translanguager (TLER)	0 = Student, 1 = Teacher, 2 = Both teacher and student, 3 = 2+ Students
Grouping (GROUP)	0 = Whole class, 1 = Small group, 2 = Pair, 3 = Alone
Teacher response (TRESP)	0 = Prohibiting, 1 = Permitting, 2 = Promoting
Grade (GRADE)	0–7
Transcription (TRAN)	
Problem Type (PROB)	Word problem = 1, Arithmetic Calculation = 2, Geometry = 3, Measurement = 4, Comparing Numbers = 5, Counting = 6
School Language Policy (LPSCHL)	0 = Strict, 1 = Flexible
Teacher Language Policy (LPTEACH)	0 = Strict, 1 = Flexible
Student Spanish Language Proficiency (SSLP)	1 = Intermediate, 2 = Advanced, 3 = Distinguished, 4 = Superior
Student English Language Proficiency (SELP)	1 = Intermediate, 2 = Advanced, 3 = Distinguished, 4 = Superior

NOTE

This research was funded, in part, by a grant from the U.S. Department of Education, Office of English Language Acquisition, National Professional Development Program, Grant #T365Z170070. MALLI is a professional development program that works to integrate mathematics, language, and literacy in dual language settings. It focuses on developing discourse, literacy and vocabulary strategies during mathematics instruction.

REFERENCES

Adler, J. (1998). A language of teaching dilemmas: Unlocking the complex multilingual secondary mathematics classroom. *For the Learning of Mathematics, 18*(1), 24–33.

Alvarez, L. (2012). Reconsidering academic language in practice: The demands of Spanish expository reading and students' bilingual resources. *Bilingual Research Journal, 35*(1), 32–52.

Borg, S. (2003). Teacher cognition in language teaching: A review of research on what language teachers think, know, believe, and do. *Language Teaching, 36*, 81–109.

Bose, A., & Choudhury, M. (2010). Language negotiation in a multilingual mathematics classroom: An analysis. In L. Sparrow, B. Kissane & C. Hurst (Eds.), *Proceedings of the 33rd Conference of the Mathematics Education Research Group of Australasia* (pp. 93–100). MERGA.

Brevik, L. M., & Rindal, U. (2020). Language use in the classroom: Balancing target language exposure with the need for other languages. *TESOL Quarterly, 54*(4), 925–953.

Brown, T. (2002). *Mathematics education and language: Interpreting hermeneutics and poststructuralism.* Kluwer Academic Publishers.

Brown, C. A., & Cooney, T. J. (1982). Research on teacher education: A philosophical orientation. *Journal of Research and Development in Education, 15(4)*, 13–18.

Canagarajah, S. (2011). Codemeshing in academic writing: Identifying teachable strategies of translanguaging. *The Modern Language Journal*, 95, 401–417.

Celedón-Pattichis, S., Musanti, S. I., & Marshall, M. E. (2010). Bilingual elementary teachers' reflections on using students' native language and culture to teach mathematics. In *Mathematics Teaching and Learning in K–12* (pp. 7–24). Palgrave Macmillan.

Cenoz, J. (2017). Translanguaging in school contexts: International perspectives. *Journal of Language, Identity and Education, 16*(4), 193–198.

Civil, M. (2010). A survey of research on the mathematics teaching and learning of immigrant students. In *Proceedings of the sixth congress of the European Society for Research in Mathematics Education* (pp. 1443–1452). *CERME 6–WORKING GROUP 8*, 1443.

Creese, A., & Blackledge, A. (2010). Translanguaging in the bilingual classroom: A pedagogy for learning and teaching? *The Modern Language Journal, 94*(1), 103–115.

de Certeau, M. (1984). *The practice of everyday life.* University of California Press.

Esquinca, A., Araujo, B., & De la Piedra, M. T. (2014). Meaning making and translanguaging in a two-way dual language program on the US-Mexico border. *Bilingual Research Journal, 37*(2), 164–181.

Flores, B. (2001). Bilingual education teachers' beliefs and their relation to self-reported practices. *Bilingual Research Journal, 25*(3), 275–299.

Forman, E. (1996). Learning mathematics as participation in classroom practice: Implications of sociocultural theory for educational reform. In L. Steffe, P. Nesher, P. Cobb, G. Goldin, & B. Greer (Eds.), *Theories of mathematical learning* (pp. 115–130). Lawrence Erlbaum Associates.

García, O. (2013). Countering the dual: Transglossia, dynamic bilingualism, and translanguaging in education. In R. Rubdy & L. Alsagoff (Eds.), *The global-local interface and hybridity* (pp. 100–118). Multilingual Matters.

García, O., & Kleyn, T. (2016). Translanguaging theory in education. In O. García & T. Kleyn (Eds.), *Translanguaging with multilingual students: Learning from classroom moments* (pp. 34–54). Routledge.

García, O., & Sylvan, C. E. (2011). Pedagogies and practices in classrooms: Singularities in pluralities. *Modern Language Journal, 95*(3), 385–400.

García, O., & Wei, L. (2015). Translanguaging, bilingualism, and bilingual education. *The handbook of bilingual and multilingual education, 223*, 240.

Gee, J. (1996). *Social linguistics and literacies: Ideology in discourses* (3rd ed.). Falmer.

Giles, H., & Ogay, T. (2007). Communication accommodation theory. In B. B. Whaley & W. Samter (Eds.), *Explaining communication: Contemporary theories and exemplars* (pp. 325–344). Lawrence Erlbaum Associates.

Gort, M., & Pontier, R. W. (2013). Exploring bilingual pedagogies in dual language preschool classrooms. *Language and Education, 27*, 223–245.

Gort, M., & Sembiante, S. F. (2015). Navigating hybridized language learning spaces through translanguaging pedagogy: Dual language pre-school teachers' languaging practices in support of emergent bilingual children's performance of academic discourse. *International Multilingual Research Journal, 9*, 7–25.

Halliday, M. A. K., & Martin, J. R. (1993). *Writing science: Literacy and discursive power.* University of Pittsburgh Press.

Howard, E. R., Sugarman, J., Perdomo, M., & Adger, C. T. (2005). Two-way immersion education: The basics. *The Education Alliance at Brown University.* http://www.cal.org/twi/toolkit/PI/Basics_Eng.pdf

King, N., & Ridley, J. (2019). A Bakhtinian take on languaging in a dual language immersion classroom. *System, 80*, 14–26.

Kvale, S. (1996). *Interviews: An introduction to qualitative research interviewing.* Sage.

Lemke, J. (1990). *Talking science: Language, learning and values.* Ablex.

Lin, A. (2013). Classroom code-switching: Three decades of research. *Applied Linguistics Review, 4*(1), 195–218.

Lindholm-Leary, K. (2006). What are the most effective kinds of programs for English language learners? In E. Hamayan & R. Freeman (Eds.), *English language learners at school* (pp. 64–85). Caslon.

Lindholm-Leary, K. (2013). Education: Dual language instruction in the United States. *Americas Quarterly, 7*(4), 97.

Macaro, E. (2005). Code-switching in the L2 classroom: A communication and learning strategy. In E. Llurda (Ed.), *Non-native language teachers: Perceptions, challenges and contributions to the profession* (Vol. 5, pp. 63–84). Springer Science & Business Media.

MacSwan, J. (2017). A multilingual perspective on translanguaging. *American Educational Research Journal, 54*(1), 167–201.

Makoni, S., & Pennycook, A. (Eds.). (2007). *Disinventing and reconstituting languages* (Vol. 62). Multilingual Matters.

Martínez, R. A. (2010). Spanglish as literacy tool: Toward an understanding of the potential role of Spanish-English code-switching in the development of academic literacy. *Research in the Teaching of English, 45*(2), 124–149.

Martínez, R. A. (2009). *Spanglish is spoken here: Making sense of Spanish–English code-switching and language ideologies in a sixth-grade English Language Arts classroom* [Unpublished doctoral dissertation]. UCLA.

Martínez, R. A., Hikida, M., & Durán, L. (2015). Unpacking ideologies of linguistic purism: How dual language teachers make sense of everyday translanguaging. *International Multilingual Research Journal, 9*(1), 26–42.

Martínez-Álvarez, P. (2017). Language multiplicity and dynamism: Emergent bilinguals taking ownership of language use in a hybrid curricular space. *International Multilingual Research Journal, 11*(4), 255–276.

McCarthy, P. A. (2018). Why children code-switch? Sociolinguistic factors. *International Journal of Language and Linguistics, 5*(3), 16–40.

Moore, P. J. (2013). An emergent perspective on the use of the first language in the English-as-a foreign-language classroom. *The Modern Language Journal, 97*(1), 239–253.

Moschkovich, J. (2007). Using two languages when learning mathematics. *Educational Studies in Mathematics, 64,* 121–144.

Myers-Scotton, C. (1995). *Social motivations for code-switching: Evidence from Africa.* Oxford University Press.

Myers-Scotton, C. (2006). *Multiple voices: An introduction to bilingualism.* Blackwell Publishing.

Nambisan, K. A. (2014). *Teachers' attitudes towards and uses of translanguaging in English language classrooms in Iowa* [Unpublished thesis]. Iowa State University Digital Repository.

Nava, N. (2009). Elementary teachers' attitudes and beliefs towards their students' use of code-switching in South Texas. *Lenguaje, 37*(1), 135–158.

Otheguy, R., García, O., & Reid, W. (2015). Clarifying translanguaging and deconstructing named languages: A perspective from linguistics. *Applied Linguistics Review, 6*(3), 281–307.

Palmer, D. K., Martínez, R. A., Mateus, S. G., & Henderson, K. (2014). Reframing the debate on language separation: Toward a vision for translanguaging pedagogies in the dual language classroom. *The Modern Language Journal, 98*(3), 757–772.

Patthoff, A., Castillo, J., & Treviño, A. (2021). Dual-language teachers' use of technology to facilitate mathematical discourse. *Computers in the Schools, 38*(3), 161–188.

Patton, M. (2002). *Qualitative research and evaluation methods* (3rd ed.). Sage Publications.

Pimm, D. (1987). *Speaking mathematically: Communication in Mathematics classrooms.* Routledge.

Planas, N. (2018). Language as resource: A key notion for understanding the complexity of mathematics learning. *Educational Studies in Mathematics, 98*(3), 215–229.

Planas, N., & Civil, M. (2013). Language-as-resource and language-as-political: Tensions in the bilingual mathematics classroom. *Mathematics Education Research Journal, 25,* 361–378.

Planas, N., & Setati, M. (2009). Bilingual students using their languages in the learning of mathematics. *Mathematics Education Research Journal, 21*(3), 36–59.

Roth, W. M., & Tobin, K. (Eds.) (2007). *Science, learning, identity: Sociocultural and cultural–historical perspectives.* Sense.

Salehmohamed, A., & Rowland, T. (2014). Whole-class interactions and code-switching in secondary mathematics teaching in Mauritius. *Mathematics Education Research Journal, 31,* 431–446.

Setati, M. (2005). Learning and teaching mathematics in a primary multilingual classroom. *Journal for Research in Mathematics Education, 36*(5), 447–466.

Setati, M. (2008). Access to mathematics versus access to the language of power: The struggle in multilingual mathematics classrooms. *South African Journal of Education, 28,* 103–116.

Setati, M., & Adler, J. (2000). Between languages and discourses: Language practices in primary multilingual mathematics classrooms in South Africa. *Educational Studies in Mathematics, 43,* 243–269.

Setati, M., Adler, J. Reed, Y., & Bapoo, A. (2002). Incomplete journeys: Code-switching and other language practices in mathematics, science and English language classrooms in South Africa. *Language and Education, 16*(2), 128–149.

Skott, J. (2015). The promises, problems, and prospects of research on teachers' beliefs. *International Handbook of Research on Teachers' Beliefs, 1,* 37–54.

Solomon, Y. (2009). *Mathematical literacy: Developing identities of inclusion.* Routledge.

Somerville, J., & Faltis, C. (2019). Dual languaging as strategy and translanguaging as tactic in two-way dual language programs. *Theory Into Practice, 58*(2), 164–175.

Tavares, N. J. (2015). How strategic use of L1 in an L2-medium mathematics classroom facilitates L2 interaction and comprehension. *International Journal of Bilingual Education and Bilingualism, 18,* 319–335.

Torres-Guzmán, M. (2007). Dual language programs: Key features and results. In O. García & C. Baker (Eds.), *Bilingual education: An introductory reader* (pp. 50–63). Multilingual Matters.

U.S. Department of Education. (2019). *The condition of education.* Retrieved from http://nces.ed.gov/programs/coe/indicator_cgf.asp

Volterra, V., & Taeschner, T. (1978). The acquisition and development of language by bilingual children. *Journal of Child Language, 5,* 311–326.

Wang, L., & Kirkpatrick, A. (2013). Trilingual education in Hong Kong primary schools: A case study. *International Journal of Bilingual Education and Bilingualism, 16*(1) 100–116.

Weber, J. J. (2014). *Flexible multilingual education: Putting children's needs first.* Multilingual Matters.

Windshitl, M. (2002). Framing constructivism in practice as the negotiation of dilemmas: An analysis of the conceptual, pedagogical, cultural, and political challenges facing teachers. *Review of Educational Research,* 72, 131–175.

CHAPTER 4

THE SYMBIOTIC RELATIONSHIP BETWEEN LANGUAGE AND MATHEMATICS

Bootstrapping Writing, Discourse, and Mathematics in Dual Language Programs

Carmina Mendoza
Santa Clara University

ABSTRACT

Recently, especially in California, there has been a renowned commitment to support our English language learners (California Department of Education, 2017). With the adoption of the English Learner Roadmap, California seeks to find instructional support to welcome, understand, and educate our diverse student population of emergent bilingual learners (California Department of Education, 2017). One way to support multilingual learners is to bootstrap the development of English and home language skills with the acquisition of academic proficiency. By using home language as a leverage, both

Mathematics Instruction in Dual Language Classrooms, pages 63–78

content and language skills can be strengthened. In this chapter, I explain the structural pillars of effective dual language programs and how they support mathematical instruction within the context of the "new way" of teaching mathematics in TK–12. The new approaches to mathematics teaching include common core standards, different levels of mathematical understandings, and math coherence across standards and skills. I also address the language and literacy connection in mathematics by presenting the salient literature that attests to the value of writing and discourse in scaffolding mathematical understandings, concepts, and skills. Finally, research shows that the dual language program is the optimal environment for mathematics learning because the use of home language allows students to access a broader linguistic repertoire to express mathematical concepts and reasoning (Moschkovich, 2007c). By supporting mathematical proficiency through dual language programs, we are contributing to a more equitable educational pathway for our multilingual learners.

DUAL LANGUAGE IMMERSION—A MATTER OF MEANINGFUL ACCESS TO EDUCATION

Because the EL Roadmap (ELR) developed from the latest research-based theoretical frameworks (California Department of Education, 2017), I would like to adopt it as one of the most current guidelines to support multilingual learners and unpack what a sound bilingual program involves. The ELR's vision promotes the idea that in order for multilingual learners to develop English language skills, academic mastery, and multiple languages proficiency, we must support our multilingual learners to *fully participate and gain meaningful access in an education setting* (California Department of Education, 2017). Indeed, bilingual education in the United States emerged as an issue of equal access to education as defined by the ELR vision (Spitzer, 2019).

Bilingual Education is anchored on the idea that home language instruction along with English language development supports multilingual students to fully participate in the classroom and gain meaningful access to education (Moran, 2007; Olah, 2023). In 1974, the *Lau v. Nichols* decision concluded that schools could not claim they were giving non-native English speakers the same access to education as native English speakers if education was provided in a language they did not understand (Olah, 2023; Spitzer, 2019). The Supreme Court concluded that the "school district's exclusive reliance on English-language instruction wrongfully excluded non-English-speaking children from access to the curriculum in violation of Title VI. 2" (Moran, 2007, p. 4). *Lau v Nichols* paved the road to ensure that our multilingual learners were entitled to "a meaningful opportunity to participate in the educational program" (Olah, 2023, para.1). Rachel

Moran (2007) concludes that *Lau v Nichols* "effectively added English language learners (ELLs) to the growing list of U.S. citizens to benefit from the civil rights movement" (p. 4). Bilingual education is one way to ensure that the civil rights of our multilingual learners are protected by including home language as the language of instruction.

THE FOUR PILLARS OF DUAL LANGUAGE EDUCATION

The focus of this chapter is not to present an exhaustive overview of the literature on bilingual education models. Rather, I would like to explore the idea that dual language immersion programs provide an effective structure to promote equal access to academic content while developing language skills in two languages—the home language and the target language. Donna Christian (2016, p. 1) provides a typology of bilingual programs as follows:

- *Bilingual Programs*: Students share the same language background and are learning English as an additional language.
- *Foreign/World Language Immersion Programs or One-Way Immersion*: Students are mostly native English speakers who are learning another language.
- *Heritage Language Immersion Programs*: Students learn the language of their shared background.
- *Two-Way Immersion Programs*: Native speakers of the target language and native English speakers learn each other's language and are represented in roughly equal proportions in a classroom.

In this chapter, I am adopting my own working definition of dual language immersion programs as any program in which students from two different linguistic backgrounds are immersed in meaningful language learning and academic engagement by using the two languages represented by the participating students as learning tools. Language is a tool to access content either by communicating orally or in written form. In the context of a lesson, students are actively engaged in interpretive (listening and reading), interpersonal, and presentational language (speaking and writing) formats.. Therefore, language becomes a tool or scaffold for accessing mathematical content. When we use the whole language repertoire of the students (home and target language), students have more *tools* to make sense of content, express thinking processes, at the same time that they develop language (Moschkovich, 2007c)

As an educational model, bilingual programs, especially dual language immersion have taken different forms with diverse results. Some researchers have documented the academic benefits of bilingual education (National

Academies of Sciences, Engineering, and Medicine, 2017; Serafini et al., 2022). Controlling for several student background variables, Serafini and her colleagues (2022) identified that bilingual instruction was associated with faster English language acquisition and higher GPAs. In Serafini's study, the role of home language support had a significant impact on the positive outcomes. However, other researchers have argued that some bilingual programs in the United States, rather than promoting equality, have perpetuated *neoliberal ideas* that have negatively impacted English language learners, while enhancing the learning of English native speakers, often representatives of the mainstream society (Freire et al., 2022).

Freire (2022) and his team analyzed Utah's dual language bilingual education, a 50–50 program. Using critical discourse analysis of policy documents from states' websites across the United States, this study found that Delaware, Georgia, and Wyoming emulated Utah's model, which showed discursive gentrification that benefited White English speakers (Freire et al., 2022). Discursive gentrification is defined as the "invasion of privileged students" into bilingual programs (Freire et al., 2022, p. 39). Bilingual programs can produce inequality if native English speakers receive more support and encouragement, compared to the support that the non-native English speakers receive. This is the result of policies that promote dominant language practices such as language isolation, instead of more inclusive practices such as translanguaging and code-switching. Translanguaging and code-switching allow students to express their ideas using a linguistic repertoire that may include multiple languages (Krauss et al., 2022). According to Garcia (2009), "Translanguaging is the act performed by bilinguals of accessing different linguistic features or various modes of what are described as autonomous languages, in order to maximize communicative potential" (p. 140). Translanguaging allows emergent bilingual students to access a broader collection of language resources to express ideas and demonstrate their knowledge. Freire's study reminds us of the importance of the interconnectedness of policy, research, and instruction in education. Without an equitable guiding policy, research and instruction cannot be implemented effectively or equitably.

What makes an effective bilingual program can be up to debate. However, many dual language immersion programs in the United States have been developed and implemented the following seven research-based strands outlined in the *Guiding Principles for Dual Language Education*:

1. Program Structure
2. Curriculum
3. Instruction
4. Assessment and Accountability
5. Staff Quality and Professional Development

6. Family and Community
7. Support and Resources (Howard et al., 2018)

These seven strands aim to support students within a fully integrated program that promotes (a) bilingualism, (b) biliteracy, and (c) sociocultural competencies (Howard et al., 2018). These three areas have been considered *the three pillars of dual language education* (Howard et al., 2018). These three pillars are aligned to the EL Roadmap in regards to the ability of a bilingual program to embrace an asset-oriented, needs-based approach to instruction, as well as providing students with meaningful access to high quality instruction. First, bilingualism and biliteracy enables students to reach that meaningful access to education via high quality instruction as the content can be delivered or supported in languages that students are most familiar with. Second, the sociocultural competencies pillar supports the notion that our students in dual language programs come to school with a lot of experiences that constitute "assets" that can support an ongoing learning experience in the classroom. Paying attention to sociocultural competencies also supports students' specific needs that can affirm their culture and identities by developing the home language as they acquire English skills.

A solid dual language immersion program should develop bilingualism (listening and speaking skills in two languages), biliteracy (reading and writing skills in two languages), and appreciation and willingness to learn about multiple cultures—including our own. Recently, a new pillar has been added (Cervantes-Soon et al., 2017). This new pillar aims to minimize the negative effects of bilingual programs (Freire et al., 2022). Cervantes-Soon and her colleagues (2017) explored several inequalities at both the structural and classroom levels within dual language immersion programs. They suggested that these inequalities could be addressed by adding another pillar—*the critical consciousness* (Cervantes-Soon et al., 2017). This pillar focuses on empowering students to develop a critical mindset where they can identify injustice and actively seek ways to denounce it and reverse it (Cervantes et al., 2017). In order to implement critical consciousness, participants in a dual language immersion should reframe their relationships in a humanistic way by identifying and honoring their place in this world and their role in oppression dynamics (Dorner et al., 2019). Further, classrooms should be safe spaces to engage in ongoing dialogues about power structures and demystifying stereotypes (Dorner et al., 2019). This fourth pillar invites us to restructure our dual language immersion programs as an opportunity to implement liberating discourses.

Analyzing these four pillars, I conclude that dual language programs aim to give meaningful access to education to multilingual learners by developing literacy and language skills in both languages (Howard et al., 2018). These skills should support students in their academic development by equipping them with the tools to meet academic standards (California Department of Education, 2017). This is usually done by integrating language development objectives with academic objectives (Kumar, 2018) and the use of thematic units and problem-based learning (Howard et al., 2018). Further, dual language immersion programs should not only develop cultural appreciation and competence (Howard et al., 2018), but also critical consciousness (Cervantes-Soon et al., 2017; Dorner et al., 2019). Math, as a subject, can be used to provide instruction that supports these four dual language education pillars.

THE TEACHING OF MATHEMATICS—REASSESSING AND INNOVATING

Just like bilingual education, mathematics education has been subjected to multiple policies that have aimed to improve U.S. students' performance in mathematics (OECD, 2023). According to PISA (Program for International Student Assessment), a measure used by countries under the Organisation for Economic Co-operation and Development (OECD), the United States has consistently scored below the OECD average in mathematics for the last 20 years (DeSilver, 2017). In 2005, the American Institutes for Research prepared for the U.S. Department of Education the *Reassessing U.S. International Mathematics Performance: New Findings From the 2003 TIMSS and PISA Report* (Ginsburg et al., 2005). This report aimed to identify the factors that influenced the 2003 low scores in mathematics for U.S. students. The report found that rigor issues, unequal distribution of time allocated to different content areas of mathematics, and lack of national standards for mathematical instruction were factors that were related to the low math scores for U.S. students (Ginsburg et al., 2005). Since then, major policies and instructional approaches in mathematics have been introduced to improve the performance of U.S. students. These include the Common Core Standards (Porter et al., 2011), mathematical practices (Schoenfeld, 2020), the intentional inclusion of diverse types of mathematical knowledge—conceptual, procedural, application (Star & Stylianides, 2013), and the Coherence Map, which provides a roadmap of vertical and horizontal alignments for the Common Core Standards and mathematical skills (Achieve the Core, 2024).

The Common Core Standards include both content standards and mathematical practices that provide students the knowledge and skills they need at each grade level (National Council for the Teaching of Mathematics, 2017). The Mathematical Practices address more than knowledge. It is a set of eight practices that aim to promote the mindset of real-life mathematicians in TK–12 students. These eight practices are based on a combination of the

National Council for the Teaching of Mathematics' practices regarding process standards and the strands of mathematical proficiency recommended by the National Research Council (California Department of Education, 2023). Common Core Standards and the Mathematical Practices are an attempt to provide the rigorous, balanced instruction, under national standards, that were needed in mathematical instruction in the United States.

LANGUAGE AND MATHEMATICS: A SYMBIOTIC INSTRUCTIONAL RELATIONSHIP

Language, literacy, and math skills have always been considered "basic skills." Literacy includes both reading and writing; while language includes interpretive, interpersonal, and presentational communications, in both oral and written form. In this chapter, I will concentrate on two aspects of literacy and language—writing and discourse. Within these concepts, vocabulary development is key as students need specific vocabulary to *write and speak math.*

Many schools emphasize English language arts and mathematics as the most important core knowledge and skills needed for success in life and career. The adoption of the Common Core Standards brought a laser-focused priority to the teacher profession to implement the English language arts and math standards (Opfer et al., 2016). Many elementary schools devote more instructional time to math and language arts than any other subject, and often choose an integrated approach to the teaching of science and social studies, which integrates both language arts and math into the other subject areas. This integration is evidenced in the fact that many schools require their teachers to have both a content and a language objective in each lesson (Kumar, 2018). This can be due in part to the fact that most standardized tests at the state level involve math and language skills.

One of the major myths in the instruction of mathematics is that math is a less language-dependent subject (Ewing, 2020). On the contrary, there is a strong connection between language and reasoning and problem-solving skills (Dale & Cuevas, 1992; Jarret, 1999). Language and math are in a symbiotic instructional relationship. When learning math, students can use their linguistic skills to enhance their mathematical understanding. When language lessons include math, the nature of the subject matter presents an opportunity for students to develop language skills as these are needed for reasoning and problem solving. With the implementation of the Common Core and Mathematical Practices, language has a central role in the mathematics classroom, especially for English language learners (Ewing, 2020; Slavit & Slavit, 2015). The concepts of writing to learn, mathematical discourse, and integrated language development are three practices that demonstrate the symbiotic instructional relationship between language and math. These instructional approaches reinforce the dual language education pillars of bilingualism and biliteracy. The two other pillars—cultural competence and critical consciousness—are also present in math under the concepts of math identity and ethnomathematics, which can be supported by appropriate mathematical discourse.

WRITING TO LEARN IN THE MATHEMATICS CLASSROOM

Writing to learn has been documented as an effective way to help students to process their mathematical knowledge (Athanases & de Oliveira, 2014; Flores & Brittain, 2004; Gibbons, 2016). Writing to learn is more than a method to document mathematical knowledge (show your answer). The concept of writing to learn involves students using writing skills to organize, analyze, and reflect on mathematical concepts (Bravo et al., this volume; Countryman, 1992). It provides an opportunity for students to deepen their understanding of mathematics and seek new perspectives (Burns, 2004; Urquhart, 2009). Pugalee (2005) found that writing specifically helps support students' mathematical reasoning and problem-solving skills. The systematic and intentional inclusion of writing activities such as learning logs, word problems, and explaining mathematical ideas and thinking-processes allow teachers to support their students' mathematical performance. The benefits of writing to learn are not limited to K–12 students. Flores and Brittain (2003) identified several benefits of writing for teachers in a mathematics methods course. Math teachers can use writing to make math relevant, to organize ideas and thinking processes, to reflect on an ongoing basis, and to address affective issues regarding difficult issues and pressures in teaching mathematics (Flores & Brittain, 2003).

MATHEMATICAL DISCOURSE: EMBEDDING VOCABULARY, MATH IDENTITY, AND ETHNOMATHEMATICS

Besides written language, oral language is also crucial in scaffolding mathematical learning. Many researchers have argued that the inclusion of mathematics discourse is crucial in supporting multilingual learners in acquiring mathematical knowledge and creating a positive learning community towards math (Anderson et al., 2011; Mitchell & Knuth, 2003; Moschkovich, 2007c; The Education Trust-West, 2018). By using mathematical discourse, students are invited to engage in discussions that not only build mathematical knowledge, but also promote critical thinking and risk-taking in learning (Hansen-Thomas, 2009; Holland et al., 2017; Joswitch, 2017; Lim et al., 2020). Mathematical discourse practices are more than asking and answering questions. Effective math discourse practices should encourage learners to talk about math in the context of everyday and professional applications of mathematics (Moschkovich, 2007b, 2007c). These practices include explaining reasoning, presenting solutions and alternative models, and listening to others' ideas (Anderson et al., 2011; Lim et al., 2020). These practices must lead to the building of mathematical knowledge through collaboration (Lim et al., 2020).

Vocabulary is an important component of effective discourse because in mathematics, students are often faced with using specialized vocabulary in the context of solving problems in everyday tasks (Moschkovich, 2007c). Discourse and writing require effective vocabulary strategies that allow students to develop a toolkit to "speak and write" math. In the specific case of Spanish, English learners can capitalize on their Spanish knowledge since many math terms are cognates with a Latin etymology. For Spanish speaking students, Spanish knowledge provides an advantage to access academic language in English (Lubliner & Hiebert, 2011). One of the best ways to make mathematical discourse inclusive is the incorporation of translanguaging (Garcia, 2009). In translanguaging, bilingual teachers mobilize diverse language resources available to students in order to help them demonstrate knowledge and express ideas (Tai, 2022). Students are able to access mathematical content as well as language skills by using a broader pool of linguistics resources from home and target languages (Moschkovich, 2007c; Tai, 2022).

Mathematical discourse is also an important element in developing a mathematical identity in students because discourse proficiency is one way to access specific social spaces. Mathematical identity refers to the degree that students see themselves as having self-efficacy in mathematics and having the skills and knowledge to interact in social spaces where mathematics is discussed (Solomon, 2008). Math identity is based on beliefs about one's self as a math learner and others' perceptions of our math abilities (REL Northwest, 2019). Students from marginalized populations often feel pressured to achieve mathematical excellence against the background of racialized experiences, which leads students to develop fragilized mathematical identities (O'McGee, 2015). Strong mathematical identities lead to the development of mathematical agency (REL Northwest, 2019). Mathematical agency refers to students' ability to freely express their mathematical identity (Aguirre et al., 2013). Learning math in a context that allows home language to be utilized as an important learning tool that can support multilingual students to adopt healthier mathematical identities, rather than fragmented ones. Tai (2022) argues that using translanguaging in mathematical discourse provides students with a bridge between students' home culture and the culture of mathematics (Tai, 2022), strengthening those mathematical identities.

Even though mathematics is often constructed as a "hard science," some have argued that it is impossible to ignore the effect of culture in mathematics (Lett, 1996; Rosa & Orey, 2013). The field of ethnomathematics emerged as the study of the relationship between culture and mathematics (D'Ambrosio, 1987). Ethnomathematics provide a framework to analyze the "types" of discourse that occur during the teaching and learning of mathematics–emic and etic constructs (Lett, 1996). Ethnomathematics explores how mathematics is practiced among different cultural groups (Gardes, 1994). Rosa and Orey (2013) proposed a pedagogical approach called *Ethnomodeling* that allows teachers to connect both the academic and cultural aspects of mathematics. The incorporation of Ethnomathematics in the dual language immersion classroom provides activities that capitalize on the cultural diversity that students bring to the classroom and how they can support mathematical understanding and reasoning. Based on the work of Lett (1996), Rosa and Orey (2013) propose a balanced approach to mathematical instruction that include both the emic and etic constructs of mathematical knowledge. Emic refers to the mathematical accounts, descriptions, and analyses that are socially constructed within a specific cultural group (Lett, 1996). Emic constructs are socially accepted in the cultural group and the validation of the emic knowledge comes from the "consensus of local people who must agree that these constructs match the shared perceptions that portray the characteristic of their culture" (Rosa & Orey, 2013, p. 69). Etic constructs, on the other hand, are accounts, descriptions, and analyses of mathematical knowledge accepted by the researchers in the mathematical field and it is validated by logical and empirical analysis (Lett, 1996). As such, an etic construct is considered precise, logical, and replicable by multiple observers-researchers (Lett, 1996). In a mathematics classroom, teachers often use sentence starters to support students to describe their thinking. In general, these sentence starters include "etic" constructs that include mathematical discourse that has been vetted by academics (e.g., volume, circular base, angle). But students could also include their own working definitions that have been adopted by their social circles. For example, a teacher could introduce the concept of volume of a cylinder by pointing out that the cylinder has a "circular base" (etics). But students could choose to say (instead of circular base), "The circle at the bottom of the tank." This could be because in their social space a cylinder shape is often found in water tanks. Or because as they are developing English skills, they may not know that the word "circular base" does refer to the "bottom of the tank" that happens to be in a circular shape. In order to promote inclusion in the mathematics classroom, both etic and emic constructs should be explored to support the development of our students' critical consciousness. Discourse activities that bring students together to explore emic and etic constructs can support the development of students questioning which etic constructs are valid and who makes those validation decisions.

MATH AND INTEGRATED LANGUAGE DEVELOPMENT

Research in math education in bilingual classrooms has paid attention to the interconnectedness of math and language development (Cervantes-Soon et al., 2017; Joswich, 2017; Mosqueda & Maldonado, 2013; National Academies of Sciences, Engineering, and Medicine, 2017; Palmer et al., 2019; The Education Trust-West, 2018). In the integrated approach to language development and content area learning, lessons have both content and language objectives (Kumar, 2018; The Education Trust-West, 2018). For example, multilingual learners learn math during language arts lessons when the teacher uses a math theme to teach a language lesson. On the other hand, students continue to develop their language skills in both home and target language when dual language immersion teachers incorporate language scaffolds during a math lesson. The "tandem" approach is an instructional approach to implement the English Language Development Standards, the Common Core State Standards for Mathematics, and the Next Generation Science Standards for California Public Schools (Lagunoff et al., 2015). In this approach math teachers with multilingual learners in their classrooms address content and literacy standards in the same lesson (Lagunoff et al., 2015). The Dual Immersion classroom has the added benefit of having students develop language skills in two different registers that can provide them with a broader linguistic repertoire to access knowledge (Garcia, 2009). Moschkovich (2007b) analyzed mathematical discourse practices of bilingual students when using their full linguistic repertoire to explain arithmetic computations. Code switching, rather than being a source of confusion, deficiency, or delay in learning, provided students with resources to communicate mathematically (Moschkovich, 2007b).

The *Guiding Principles for Dual Language Education* anchors their curricular recommendations in cross-disciplinary and project-based learning approaches (Howard et al., 2018). Cross-disciplinary approaches develop both language and literacy skills and promote content area competencies (Halvorsen et al., 2014). This approach has been documented to be especially beneficial for English language learners (National Academies of Sciences, Engineering, and Medicine, 2017).

CONCLUSION

In this chapter, I have suggested that bilingual education is anchored in the concepts of equality and equity. As an educational policy, bilingual education aims to ensure that multilingual learners in public schools have meaningful access to instruction. Supporting emergent bilinguals in the math classroom requires one to bootstrap language and mathematics by teaching academic vocabulary (Jarret, 1999), reading and understanding word problems (Bernardo, 2005), and supporting students with language production in the content area, including the use of translanguaging practices (Krause et al., 2022). Further, math assessment of emergent bilinguals should incorporate tools to allow them to express mathematical ideas independently of English language proficiency (Moschkovich, 2007a). The use of ethnomathematics frameworks can help teachers to enhance their understanding of *prior knowledge* by identifying how much mathematical knowledge their students may have under emic constructs, rather than etic ones.

Moreover, strong bilingual programs have a strong literacy and language development component, as well as a commitment to develop multicultural competence and critical consciousness in students. These pillars of dual language education are interconnected with current practices in teaching mathematics. Mathematics is no longer a procedural subject where rules and rote practice create mathematical proficiency. Today's math instruction focuses on using language as a tool that can provide effective scaffolding to access math knowledge. The use of writing and discourse in the mathematics classroom aims to create more inclusive classrooms where math is accessible to all students. Promoting mathematical discourse practices that are culturally responsive and recognize principles of ethnomathematics is a way to strengthen our students' mathematical identity and agency.

REFERENCES

Aguirre, J., Mayfield-Ingram, K., & Martin, D. (2013). *The impact of identity in K–8 mathematics: Rethinking equity-based practices.* The National Council of Teachers of Mathematics.

Anderson, N., Chapin, S., & O'Connor, C. (2011). *Classroom discussions: Seeing math discourse in action, grades K–6.* Math Solutions.

Athanases, S. Z., & de Oliveira, L. C. (2014). Scaffolding versus routine support for Latina/o youth in an urban school: Tensions in building toward disciplinary literacy. *Journal of Literacy Research, 46*(2), 263–299.

Bernardo, A. I. (2005). Language and modeling word problems in mathematics among bilinguals. *The Journal of Psychology, 139*(5), 413–425.

Burns, M. (2004). Writing in math. *Educational Leadership, 62*(2), 30–33.

California Department of Education. (2017). *English learner roadmap—English learners.* https://www.cde.ca.gov/sp/el/rm/

California Department of Education. (2023, July 25). *Standards for mathematical practice.* https://www.cde.ca.gov/be/st/ss/mathpractices.asp

Cervantes-Soon, C., Dorner, L., Palmer, D., Heiman, D., Schwerdtfeger, R., & Choi, J. (2017). Combating inequalities in two-way language immersion programs: Toward critical consciousness in bilingual education spaces. *Review of Research in Education, 41*(1), 403–427.

Christian, D. (2016). Dual language education: Current research perspectives. *International Multilingual Research Journal, 10*(1), 1–5.

Countryman, J. (1992). *Writing to learn mathematics.* Heinemann.

D'Ambrosio, U. (1987). Reflections on ethnomathemathematics. International study group on ethnomathemathematics. *Newsletter, 3,* 3–5.

Dale, T. C., & Cuevas, G. J. (1992). Integrating mathematics and language learning. In P. A. Richard-Amato & M. A. Snow (Eds.), *The multicultural classroom: Readings for content-area teachers* (pp. 330–348). Longman.

DeSilver, D. (2017). *U.S. students' academic achievement still lags that of their peers in many other countries.* Pew Research Center. https://www.pewresearch.org/short-reads/2017/02/15/u-s-students-internationally-math-science

Dorner, L., Palmer, D., Cervantes-Soon, C. (February 9, 2019). *Proposing a fourth goal for equitable dual language education: Ideological clarity for all* [Conference Presentation]. International Conference on Immersion and Dual Language Education, Charlotte, NC.

Ewing, J. (2020). *Math & ELLs: As easy as uno, dos, tres.* Rowman & Littlefield Publishers.

Flores, A., & Brittain, C. M. (2003). Writing to reflect in a mathematics methods course. *Teaching Children Mathematics, 10*(2), 112–118.

Flores, A., & Brittain, C. M. (2004). Writing for an audience in a mathematics methods course. *Teaching Children Mathematics, 10*(9), 480–486.

Freire, J. A., Gambrell, J., Kasun, G. S., Dorner, L., & Cervantes-Soon, C. (2022). The expropriation of dual language bilingual education: deconstructing neoliberalism, whitestreaming, and English-hegemony. *International Multilingual Research Journal, 16*(1), 27–46.

García, O. (2009). Education, multilingualism and translanguaging in the 21st century. In T. Skutnabb-Kangas, R. Phillipson, A. K. Mohanty, & M. Panda (Eds.), *Social justice through multilingual education* (pp. 140–158). Multilingual Matters. https://doi.org/10.21832/9781847691910-011

Gardes, P. (1994). Reflections on ethnomathematics. *For the Learning of Mathematics, 14*(2), 19–22. https://www.jstor.org/stable/40248110

Gibbons, P. (2016). *English learners, academic literacy, and thinking. Literacy in the challenge zone.* Heinemann.

Ginsburg, A., Geneise C., Leinwand, S., Noell, J., & Pollock, E. (2005). *Reassessing U.S. international mathematics performance: New findings from the 2003 TIMSS and PISA.* American Institutes for Research.

Halvorsen, A., Duke, N. K., Brugar, K. A., Block, M. K., Strachan, S. L., Berka, M. B., & Brown, J. M. (2014). Narrowing the achievement gap in second-grade social studies and content area literacy: The promise of a project-based approach. *Theory and Research in Social Education, 40,* 198–229.

Hansen-Thomas, H. (2009). Reform-oriented mathematics in three 6th grade classes: How teachers draw in ELLs to academic discourse. *Journal of Language, Identity & Education, 8*(2–3), 88–106.

Holland, W. B., Palacios, N. A., Merritt, E.g., & Rimm-Kaufman, S. E. (2017). Scaffolding English language learners' mathematical talk in the context of calendar math. *The Journal of Educational Research, 110*(2), 199–208. https://doi.org/10.1080/00220671.2015.1075187

Howard, E. R., Lindholm-Leary, K. J., Rogers, D., Olague, N., Medina, J., Kennedy, B., Sugarman, J., & Christian, D. (2018). *Guiding principles for dual language education* (3rd ed.). Center for Applied Linguistics.

Jarret, D. (1999). *The inclusive classroom: Teaching mathematics and science to English language learners—It's just good teaching.* Northwest Regional Educational Laboratory.

Joswick, C. D. (2017). *Investigating the relationship between classroom discourse and concept development in geometry learning* [Doctoral dissertation, Ohio State University]. OhioLINK Electronic Theses and Dissertations Center. http://rave.ohiolink.edu/etdc/view?acc_num=osu1500237810722885

Krauss, G., Adams-Corral, M., & Maldonado Rodriguez, L. A. (2022). Developing awareness around language practices in the elementary bilingual mathematics classroom. *Journal of Urban Mathematics Education, 15*(2), 8–40.

Kumar, S. (2018, October 16). *Using content and language objectives to help all students in their learning.* Peers and Pedagogy. https://achievethecore.org/peersandpedagogy/using-content-and-language-objectives-to-help-all-students-in-their-learning/

Lagunoff, R., Spycher, P., Linquanti, R., Carroll, C., & DiRanna, K. (2015). *Integrating the CA ELD standards into K–12 mathematics and science teaching and learning.* WestEd.

Lett, J. (1996). Emic-etic distinctions. In D. Levinson & M. Ember (Eds.), *Encyclopedia of cultural anthropology* (pp. 382–383). Henry Holt and Company.

Lim, W., Lee, J. E., Tyson, K., Kim, H-J., & Kim, J. (2020). An integral part of facilitating mathematical discussions: Follow-up questioning. *International Journal of Science and Mathematics Education, 18,* 377–398. https://doi.org/10.1007/s10763-019-09966-3

Lubliner, S., & Hiebert, E. H. (2011). An analysis of English–Spanish cognates as a source of general academic language. *Bilingual Research Journal, 34*(1), 76–93.

Mitchell J. N., & Knuth, E. J. (2003). A study of whole classroom mathematical discourse and teacher change. *Cognition and Instruction, 21*(2), 175–207. https://doi.org/10.1207/S1532690XCI2102_03

Moran, R. (2007). Undone by law: The uncertain legacy of *Lau v. Nichols. Berkeley La Raza Journal, 16*(1), 1–10. https://doi.org/10.15779/Z38J94M

Moschkovich, J. (2007a). Beyond words to mathematical content: Assessing English learners in the mathematics classroom. In A. Schoenfeld (Ed.), *Assessing mathematical proficiency* (pp. 345–352). Cambridge University Press.

Moschkovich, J. (2007b). Examining mathematical discourse practices. *For the Learning of Mathematics, 27*(1), 24–30. http://www.jstor.org/stable/40248556

Moschkovich, J. (2007c). Using two languages when learning mathematics. *Educational Studies in Mathematics, 64*, 121–144. https://doi.org/10.1007/s10649-005-9005-1

Mosqueda, E., & Maldonado, S. (2013). The effects of English language proficiency and curricular pathways: Latina/os' mathematics achievement in secondary schools. *Equity & Excellence in Education, 46*(2), 202–219.

National Academies of Sciences, Engineering, and Medicine. (2017). *Promoting the educational success of children and youth learning English: Promising futures.* The National Academies Press. https://doi.org/10.17226/24677

National Council for the Teaching of Mathematics. (n.d.). *Preparing America's students for college and career: Common core standards in mathematics.* https://www.nctm.org/ccssm/

Organisation for Economic Co-operation and Development. (2023). *Mathematics performance (PISA).* https://doi.org/10.1787/04711c74-en

Olah, L. N. (2023). *Every teacher a language teacher.* Penn GSE. https://www.gse.upenn.edu/review/feature/olah

O'McGee, E. (2015). Robust and fragile mathematical identities: A framework for exploring racialized experiences and high achievement among Black college students. *Journal for Research in Mathematics Education, 46*(5), 599–625.

Opfer, D., Kaufman, J. H., & Thompson, L. E. (2016). *Implementation of K–12 state standards for mathematics and English language arts and literacy: Findings from the American teacher panel.* RAND Corporation. https://www.rand.org/pubs/research_reports/RR1529-1.html

Palmer, D., Cervantes-Soon, C., Dorner, L., & Heiman, D. (2019). Bilingualism, biliteracy, biculturalism and critical consciousness for all: Proposing a fourth fundamental principle for two-way dual language education. *Theory Into Practice, 58*(2), 121–133. https://doi.org/10.1080/00405841.2019.1569376

Porter, A., McMaken, J., Hwang, J., & Yang, R. (2011). Common core standards: The new U.S. intended curriculum. *Educational Researcher, 40*(3), 103–116. https://doi.org/10.3102/0013189X11405038

Pugalee, D. (2005). *Writing to develop mathematical understanding.* Christopher-Gordon.

REL Northwest. (August 13, 2019). *Promoting a positive math identity: The importance of math identity for math success* [Conference Presentation]. Idaho Council of Teachers of Mathematics conference.

Rosa, M., & Orey, D. C. (2013). Ethnomodeling as a research theoretical framework on ethnomathematics and mathematical modeling. *Journal of Urban Mathematics Education, 6*(2), 62–80.

Schoenfeld, A. H. (2020). Mathematical practices, in theory and practice. *ZDM Mathematics Education,* 52, 1163–1175. https://doi.org/10.1007/s11858-020-01162-w

Serafini, E. J., Rozell, N., & Winsler, A. (2022). Academic and English language outcomes for DLLs as a function of school bilingual education model: The role of two-way immersion and home language support. *International Journal of Bilingual Education and Bilingualism, 25*(2), 552–570. http://doi.org/10.1080/13670050.2019.1707477

Slavit, D., & Ernst-Slavit, G. (2007). Teaching mathematics and English to English language learners simultaneously. *Middle School Journal, 39*(2), 4–11. https://doi.org/10.1080/00940771.2007.11461618

Solomon, Y. (2008). *Mathematical literacy: Developing identities of inclusion.* Routledge.

Spitzer, E. (2019). Lau v. Nichols*: Are schools required to provide bilingual instruction?* ThoughtCo. https://www.thoughtco.com/lau-v-nichols-case-4171298

Star, J. R., & Stylianides, G. J. (2013). Procedural and conceptual knowledge: Exploring the gap between knowledge type and knowledge quality. *Canadian Journal of Science, Mathematics, and Technology Education, 13,* 169–181. https://doi.org/10.1080/14926156.2013.784828

Student Achievement Partners. (n.d.). *Coherence map.* https://tools.achievethecore.org/coherence-map/

Tai, K. W. H. (2022). Translanguaging as inclusive pedagogical practices in English-medium instruction science and mathematics classrooms for linguistically and culturally diverse students. *Research in Science Education, 52,* 975–1012. https://doi.org/10.1007/s11165-021-10018-6

The Education Trust-West. (2018). *Unlocking learning II: Math as a lever for English learner equity.* https://west.edtrust.org/resource/unlocking-learning-ii-using-math-lever-english-learner-equity/

Urquhart, V. (2009). Using writing in mathematics to deepen student learning [White Paper]. McRel International. https://files.eric.ed.gov/fulltext/ED544239.pdf

PART II

SUPPORTING DUAL LANGUAGE TEACHERS WITH MATHEMATICS INSTRUCTION

CHAPTER 5

TEACHER BELIEFS TOWARD BILINGUAL INSTRUCTION IN MATHEMATICS

Eduardo Mosqueda
University of California Santa Cruz

Rachael Dektor
University of California Santa Cruz

Stephanie E. Hertel
University of California Santa Cruz

This chapter investigates the development of preservice teachers' (PSTs) beliefs following their participation in a "practice-based" teacher education program focused on the integration of language and literacy-centered pedagogical approaches for bilingual mathematics instruction within elementary-level dual language programs (DLP). Data for this study was drawn from the Mathematics Language and Literacy Integration (MALLI) project, which was designed to help PSTs develop the dispositions, knowledge, and skills needed to effectively support emergent bilingual (EB) students.

Mathematics Instruction in Dual Language Classrooms, pages 81–96

The primary goals of the MALLI project were to provide PSTs with access to mathematics methods instructors to model the integration of language and literacy practices in bilingual mathematics contexts as well as guidance from their mentor or cooperating teacher (CT) on the implementation of such strategies during their practicum experience. MALLI sought to enhance PSTs' development of beliefs and dispositions regarding their own efficacy in teaching mathematics in DLP that improved emerging bilingual (EB) students' mathematical reasoning and understanding using three broad strategies: (a) a focus on language and literacy practices defined and presented through a mathematics methods course with an emphasis on mathematical literacy, vocabulary and discourse; (b) access to a mentor or CT; and (c) peer analysis and feedback of video recorded lessons through a lesson study framework (see Patthoff, this volume). Taken together, these three elements of the MALLI project offered PSTs a novel approach for integrating language, literacy, and mathematics instruction in a DLP. Changing PST beliefs does not necessarily suggest that their instructional *practices* will change, but changing beliefs regarding one's efficacy and understanding is an important first step in any curricular or instructional innovation.

Teacher beliefs are an important area of study given their potential to influence teachers' classroom decisions and practices (Pajares, 1992). Research has shown that beliefs can function as a filter through which people experience the world, meaning that beliefs provide a type of lens through which the holder interprets experiences (Pajares, 1992). People often hold beliefs about the world that are largely shaped by past experiences such as being a student, significant events, and interactions with influential people (Beach, 1994; Johnson, 1994). Teachers tend to hold beliefs about different aspects of their educational contexts. For example, teachers hold beliefs about the value of teaching a particular curriculum or using a specific pedagogical approach (Levitt, 2001; Savasci & Berlin, 2012), they have beliefs about how their students learn (Polat et al., 2019), and they have self-efficacy beliefs around how they learn new content or material and their ability to teach (Bandura, 1993). Such beliefs can shape teachers' instructional decisions in the classroom.

Although beliefs can be difficult to change because they are often characterized as a stagnant construct, they are not immutable and are often shaped by an external influence (Beach, 1994; Borg, 2011; Johnson, 1994). Hansson and Wassermann (2002) have posited a *revision* as a type of belief change. The process of "revising" existing beliefs implies beliefs are constantly in a state of development. In other words, experiences and exposure to various external stimuli are constantly interacting with existing beliefs, thus providing opportunities for beliefs to continue developing. Such attention to external stimuli (Hansson & Wassermann, 2002; Schwitzgebel, 2002) and the social processes in belief changes (Mead,

1934) point to the importance of examining the development of beliefs as an ongoing, iterative process.

Current research examining teacher beliefs in mathematics content areas suggests that PSTs tend to hold beliefs about mathematics instruction based on their own experiences as learners. This can, in turn, influence their instructional practices and how they frame questions for students (Kajander, 2007). While teacher education programs intend to improve in-service teacher performance, there is evidence that PSTs' practices and beliefs are shaped by their mentor teachers. Despite this cycle, evidence indicates that courses with a "practicum-based component" can contribute to the development of beliefs. For instance, Bahr and Monroe (2008) collected pre- and post-survey data which indicated that participation in a math methods course and practicum-based component contributed to a measurable change in PSTs' beliefs about mathematics instruction. Therefore, PST participation in the MALLI projects' collaborative professional development (PD) model that included a practicum-type component would contribute to the development of PSTs' beliefs about their perceived self-efficacy for integrating language and literacy strategies with mathematics in a bilingual mathematics context.

Research Questions

1. What are preservice teachers' beliefs about the integration of literacy practices in a DLP?
2. Does participation in a practicum based program support the development of PST beliefs about the integration of literacy practices in dual language mathematics contexts?

LITERATURE REVIEW

Theoretical Framework for Understanding the Development of Teachers' Beliefs

This work is informed by a theoretical perspective on the nature of "beliefs." Beliefs are explained as a cognitive process predominantly shaped by experiences that take place in social contexts. In essence, the development of beliefs involves the ideas and viewpoints individuals construct in response to the external world and the evidence available to them. Such beliefs do not emerge in isolation but are subject to influence from other individuals, and can be molded through engagement in social processes, communities, and various groups (Hume, 2007; Loeb, 1990; Miller, 1973).

As a result, preservice teachers are likely to possess beliefs that were formed based on their personal experiences as students, information received from mentors and other educators, and from their day-to-day experiences.

The theoretical foundation of this study aligns with Mead's belief theory, as it places emphasis on the influence of social practices in the development of beliefs (Miller, 1973). Mead's framework for understanding belief formation incorporates both cognitive faculties and the external world. Mead underscores the crucial role of human participation in social processes and interactions, asserting that consciousness and rationality evolve through societal engagement. Rather than adopting an individualistic standpoint on the interpretation of group dynamics, culture, or physical environments, Mead's theory of self-development revolves around symbolic interaction such that role-taking and the utilization of linguistic gestures are essential for individual growth (Miller, 1973). Through active participation in communal activities or social processes, individuals take on the roles of others, often referred to as the "generalized other." This process of role-taking and engagement contributes to the internalization of significant symbols and the communication processes that shape the development of the mind (Mead, 1934).

Central to Mead's theory of belief formation is the concept of introspection. He describes introspection as a private, subjective process accessible only to the individual to whom it pertains. Mead terms this type of introspection "attitudes," which serve as the foundation for behaviors. Consequently, according to Mead, observable behaviors originate within the individual. However, Mead (1934) acknowledges that the social aspects and observable actions or behaviors are equally significant, and the exploration of an individual's inner workings begins with an outward examination of social elements and observable behaviors. In other words, understanding inner aspects commences by examining the social components and observable behaviors to uncover the root attitude or introspection.

The interplay between the external world and an individual's internal cognitive processes is pivotal in comprehending symbolic interactionism and Mead's perspective on learning and participation. Mead posits that all artifacts possess subjective meanings, with individuals reflecting upon and interpreting these artifacts to derive meaning. The ways in which individuals participate in social processes and react to artifacts influence how they interpret and derive meaning from them. Meaning-making does not involve mere imitation of others; rather, it is a dynamic process of action and reaction (Biesta & Tröhler, 2016). Individuals are continuously engaged in the process of meaning-making, responding to the external world while constructing meaning within their own cognitive realms (Biesta & Tröhler, 2016).

The construction of meaning and the interpretation of the world are products of both social interactions and an individual's cognitive processes. However, the initial formation of meaning occurs within a social context, with the interpretation of meaning taking place in an individual's mind. The dynamic interaction between these two facets is a process through which consciousness and abstract thinking evolve (Biesta & Tröhler, 2016). These processes and experiences elucidate how individuals construct meaning and interpret the world through their cognitive faculties while actively participating in social processes. Mead's theory regarding the development of consciousness deepens our understanding of the various definitions and conceptions of beliefs in educational research. Mead's theory underscores the idea that beliefs are shaped by past experiences and serve as a filter through which new experiences are perceived (Pajares, 1992). Leveraging Mead's (1934) theory of belief formation and development offers valuable insights into the ways beliefs are developed through experiences and active participation in social processes, informing the beliefs that educators hold when entering the classroom and shaping those beliefs as they engage in ongoing experiences and professional development opportunities (Enderle et al., 2014).

Integrating Language and Literacy Practices in Bilingual Mathematics Social Contexts

Current research has demonstrated the effectiveness of integrating language and literacy strategies with elementary-level in-service teachers. In this section, we discuss how the MALLI project language and literacy practices (mathematical literacy, vocabulary and discourse) can provide a social context and research-based professional learning opportunities that can help shape PSTs beliefs and dispositions about teaching in DLP. Research has shown that a focus on language and literacy practices that promote mathematical literacy, vocabulary, and discourse can improve student mathematical reasoning and understanding (see Xia, Balloffet, & Téllez, this volume). For example, research has shown that offering support to EB in mathematics, including discourse and literacy elements like mathematical storytelling, can enhance their ability to understand mathematical concepts, improve communication of their reasoning, and strengthen problem-solving skills. In a study of two bilingual Kindergarten teachers who developed stories related to student experiences that connected language and culture to promote problem-solving (Turner et al., 2009), teachers helped students learn in two languages to represent mathematical relationships and connect multiple representations, which resulted in improved sense-making, representation and communication of their reasoning on basic word problems. Celedon-Pattichis and her colleagues (2010) also found that at the end of the year, kindergarten students developed into more "competent and more confident problem-solvers but also as problem posers" (p. 40). A study of bilingual teachers participating in professional development focused on teaching mathematics using students' native language and culture to improve conceptual understanding (Celedón-Pattichis et al., 2010). The results reported by Celedón-Pattichis and her colleagues (2010) showed that bilingual teachers' integration of language and culturally-based practices to support the development of mathematical concepts in students' native language to communicate mathematical thinking, allowed teachers to see their students' capabilities to learn challenging mathematics problem-solving skills.

Other research has shown that focusing on mathematical discourse and vocabulary in a bilingual classroom can enhance understanding and reasoning, and develop their mathematics language register. In a study that emphasized mathematics discourse as a practice to promote mathematics learning of EBs in concert with linguistic support strategies, Avalos and Secada (2019) found that engaging EBs in mathematics discussions helped develop their mathematics register, which contributed to improving students' conceptual understanding (Avalos and Secada, 2019). The goal of "teaching deeply" discussed in this study involved using conceptually oriented lessons using manipulatives and other concrete representations with the expectation that students would justify their thinking and reasoning, which, in turn, prepared EBs for problem-solving discussions in a discourse community.

In another study, Rubinstein-Avila and colleagues (2015) examined the intersection of biliteracy, language use, mathematical discourse, and numeracy (or number sense). In this bilingual context, language use was approached as bidirectional so students were encouraged to develop academic language and literacy competencies in two languages. The teacher in this study provided students with "ample opportunities to use both English and Spanish to discuss their mathematical thinking, to problem solve, and to provide justifications" verbally and in writing (Rubinstein-Avila et al., 2015, p. 920). For example, to emphasize key vocabulary, the teacher employed a strategy that involved wearing signs around her neck containing important terms such as slope or "*pendiente*" (the Spanish translation for slope). In cases when key mathematical words were cognates (similar-sounding words with the same meaning in English and Spanish), the teacher explicitly pointed out this relationship to students. Another distinguished strategy to promote mathematical discourse biliteracy and binumerate development was through Chalk Talks where the teacher posed a question and "students could answer or informally write a statement and/or ask her, or each other, questions in either language" (p. 918). As a result, this study found that teachers' ability to tap into students' first language to clarify information and to encourage mathematical discourse and collaborative problem-solving, allowed students to demonstrate deeper levels of understanding in cases where they used their L1 to express what they are learning in L2.

The literature clearly shows how the integration of language and literacy practices (vocabulary, literacy and discourse strategies) in DLP contexts can enhance EB student mathematical reasoning and understanding. While the studies we reviewed incorporated two of the three language and literacy practices, our study is unique in that the MALLI intervention incorporated all three practices in tandem with PST. Below we describe how the three key elements of our intervention played a major role in influencing the beliefs of PSTs related to their perceived efficacy to integrate language and literacy pedagogical practices in bilingual mathematics classrooms to promote EB students' conceptual understanding and reasoning.

METHODS

Key Elements of the MALLI PD Intervention

The MALLI project from which the data for this chapter originates, included a comprehensive provision of professional development in mathematics and academic language for bilingual PST situated in the regions of San Jose, California, and San Antonio, Texas. The primary objective of MALLI's professional development methods was to foster the advancement of bilingualism and mathematical knowledge within the educational setting and home environment. This objective was achieved by disseminating practices and models to teachers and the parents of the students involved. In conjunction, the project facilitated the professional growth of CT who undertook the role of guiding and accommodating PST within the context of their dual language classrooms during the latter's PD experience.

MALLI's PD goals constituted developing pre-service teacher education and in-service professional development materials to enable teacher education programs and school districts to implement the model at their own sites. The methods to accomplish this goal included developing guiding principles to integrate math, language, and literacy in Spanish and English. The MALLI project also developed a model for parent engagement that assisted parents and teachers in supporting the child's language and math development (see Stohr, this volume), but this project was not connected to PST in any formal way.

LANGUAGE AND LITERACY PRACTICES: LITERACY, VOCABULARY, AND DISCOURSE

There were three focal MALLI teaching practices: (a) mathematics literacy/biliteracy, (b) mathematics vocabulary, and (c) mathematical discourse

(Bravo et al., 2022). *Mathematics literacy/biliteracy* practices include the types of reading and writing that are part of the mathematics discipline. Preservice teachers are equipped with models to teach writing during math time and can include practices such as written explanations that describe how students solved a math problem, writing math word problems for other peers to solve, and constructing tables and diagrams to explain their mathematical thinking. Similar instructional models are offered regarding reading math texts that include reading strategies (e.g., changing the rate of reading, utilizing their native language) and how to make sense of diagrams, tables, and charts, which require explicit instructional attention (Mosqueda et al., 2022).

Mathematics vocabulary draws attention to strategies that can be applied to provide access to the meanings of mathematical words. Math vocabulary includes words that are rare and most likely to be encountered within the discipline. Such words are often referred to as Tier 3 technical words due to the infrequency with which they appear in everyday contexts (Beck et al., 2002). Other mathematics vocabulary strategies include attention to cognate relationships which can provide access to the meaning of unfamiliar math terms in another language. *Mathematical discourse* examines the structure of oral or written explanations and arguments that occur within the mathematics classroom (Rumsey & Langrall, 2016) as well as the evidence that is suggested to be leveraged to support explanations and arguments in mathematics (Knudsen, Stevens, et al., 2018; Moschkovich, 2015).

Redesigned Bilingual Mathematics Methods Course

A key feature of the MALLI PD was to develop and implement a robust model for bilingual PST preparation. This model was grounded in the use of effective teaching practices that integrate the teaching of language, literacy, and mathematics instruction in both English and Spanish. To accomplish this goal, two anchor lessons were developed for each of the three MALLI literacy practices to introduce, define, and provide examples of integrating each practice into their instruction. Each anchor lesson was incorporated into the math methods course and was delivered fully in a Spanish language medium. Each anchor lesson served to define the nuances of the MALLI literacy practices as well as reinforce how to create a translanguaging context for incorporating each strategy in a mathematics lesson.

Mentorship: Preservice Teachers Were Assigned a Cooperating Teacher

Bilingual CTs mentored and PSTs in dual language classrooms during their clinical experience. The selection of CTS was predicated upon their demonstrated expertise and prior experience in guiding aspiring educators toward the adept implementation of integrated mathematical, linguistic, and literacy pedagogy. Professional development activities were facilitated by utilizing pedagogical resources, such as videos that captured CTs proficiently executing Spanish-language mathematical, linguistic, and literacy-integrated instruction. The CTs played an important role in the MALLI PD because they modeled, reinforced, and provided guidance, and feedback for the PSTs to learn to understand and implement the language and literacy strategies in their bilingual mathematics classrooms.

Lesson Study With Video Club (Examining Video With Peers)

Lesson study strategies were implemented to support teachers in the MALLI project. While traditional lesson study PD encompasses an academic year of targeted activities, reflection, and in-person support, the MALLI research team hypothesized that the effectiveness of lesson study would yield meaningful collaboration even in instances where teachers did not share the same students or space. In addition, MALLI integrated video club strategies to enhance collaboration (see Patthoff, this volume). PSTs recorded their instructional sessions, observed their own lessons and those of their peers, engaged in semi-structured discussions with their peers, exchanged insights derived from these dialogues within a smaller team, and lastly, the participating teachers shared their reflections with the entire group. The lesson study collaboration provided important opportunities for PSTs (in pairs) to meet with their peers to discuss a recorded lesson. Each pair of PSTs then provided feedback on the lesson, with a focus on integrating the MALLI language and literacy practices in bilingual mathematics lessons. During the Saturday morning PD sessions, the research team also provided feedback to teachers on their recorded lessons as a way to extend the lesson study process.

MALLI Survey Instrument

A specialized evaluation instrument was developed to correspond with and gauge the objectives of the MALLI project by Education Northwest (EdNW), who carried out both formative and summative evaluations to ensure a comprehensive analysis. The evaluation of the MALLI project provided performance feedback, informing program improvement and replication strategies.

The data collected was drawn from a pre-post survey, which included participants' reported use of integrated math/language instructional strategies and participant scores on measures of mathematical and language pedagogy knowledge. A central theme of the survey was teacher perceptions of effectiveness. The pre-post survey of participants included the percentage of program completers who rated the program as effective in preparing them to serve EB students and prepare them to integrate math, language, and literacy instruction. The first survey was administered pre-treatment to participants, and annually post-treatment until the end of the grant (11 months after the pre-survey) to all participants.

Participant Demographics

Participants in this project ($n = 89$) identified as 90% female, 9% male, and 1% unanswered. The average age of the participants was 26 years old, with a minimum age of 21, and maximum age of 53. 85% of participants were between ages 21–29 ($n = 76$), 5.6% were between ages 30–39 ($n = 5$), 4.5% were between ages 40–49 ($n = 4$), and ~1% were between ages 50–59 ($n = 1$). Ethnic data was also collected and was reported as follows: 28% Latinx/Hispanic ($n = 25$), 12.3% White ($n = 11$), 2.25% Black/African American ($n = 2$), 5.6% Asian ($n = 5$), 5.6% Multiracial ($n = 5$), and 70% as other ($n = 62$). The sample size for this study included 40 teachers that completed the pre- and post-survey.

Language and Literacy Practices Scales: Literacy, Vocabulary, and Discourse

We developed three researcher-created composite scales to capture the influence of the MALLI intervention based on pre-service teachers perceptions of the efficacy of the three language and literacy practices: The Literacy Practices Scale (LPS), Vocabulary Practices Scale (VPS), and Discourse Practices Scale (DPS) were composed of survey questions, designed to gauge teacher beliefs on how they perceive their language and literacy skills in relation to their teaching of mathematics. The composite variables included Likert scale survey questions measured by *strongly disagree, disagree, agree,* and *strongly agree.* Cronbach's alpha was calculated to measure the internal consistency (reliability) of each language and literacy scale, and we found all to be above a range ($\alpha = 0.90$), which is well above the acceptable range ($\alpha = 0.70$). The LPS included seven (7) items and had a Cronbach's alpha of 0.91. LPS consisted of items such as: "I know how to support math literacy skills in Spanish" and "I know how to design instruction that teaches bilingual students to write mathematics arguments." The VPS comprised six (6) items and had a Cronbach's alpha of 0.91. VPS included items such as: "I can explain how math vocabulary is developed in bilingual settings" and "I know how to use cognates to facilitate understanding of math concepts." Lastly, The DPS was composed of seven (7) items and had a Cronbach's alpha of 0.94. DPS included items such as: "I know how to create opportunities for students to explain and argue their math thinking" and "I can effectively leverage student funds of knowledge to create opportunities for students to talk."

RESULTS

Analysis of Growth in Beliefs About Literacy Integration in Math

The results in Table 5.1 show that the MALLI intervention had a powerful effect on teachers' beliefs about their efficacy of integrating language and literacy practices (literacy, vocabulary and discourse) in mathematics. The differences we observed in growth between PSTs' beliefs measured in the pre-survey and then the post-survey reflect roughly 2 standard deviations in growth, which marks a significant shift in positive development for the PSTs across all three Practices: literacy, vocabulary, and discourse. To examine these differences, we used repeated-measures *t*-tests (also known as paired samples *t*-tests) to assess the change in the three language and literacy practices in the survey across time or within subjects across the pre- and post-surveys. The literacy practices scale pre-survey mean for the PSTs was 2.53 ($SD = 0.45$), and the post-survey mean was 3.34 ($SD = 0.49$), and these differences were statistically significant ($t = -9.23$, $p < 0.05$). The mean score on the vocabulary practices scale, at the pre-survey, was 2.57 ($SD = 0.49$); the post-survey mean was 3.44 ($SD = 0.48$, $t = -8.28$, $p < 0.05$). The discourse practices scale mean on the pre-survey was 2.67 ($SD = 0.60$); the post-survey mean was 3.60 ($SD = .41$, $t = -8.75$, $p < 0.05$). The sizable and statistically significant growth in PST teacher beliefs across all three

TABLE 5.1 Development of Pre-Service Teacher Beliefs Toward Language and Literacy Integration in Bilingual Mathematics Classrooms

	Pre-Survey		Post-Survey		Pre to Post
	Mean	Std. Dev.	Mean	Std. Dev.	Mean Differences
Literacy Practices Scale	2.53	0.45	3.34	0.49	0.81***
Vocabulary Practices Scale	2.57	0.49	3.44	0.48	0.87***
Discourse Practices Scale	2.67	0.60	3.60	0.48	0.93***

language and literacy practices in bilingual mathematics contexts reflects the influence of the MALLI intervention.

The difference in PST beliefs about their ability to implement language and literacy strategies increased nearly two standard deviations from the mean in pre- to post-survey gains, suggesting that the MALLI intervention had a very strong positive effect on PST' beliefs about the role of language in bilingual mathematics contexts. These results provide a significant first step for teacher educators who seek to improve their PSTs' specific language and literacy strategies (Celedon-Pattichis et al., 2010; Maldonado et al., 2020; Mosqueda et al., 2022; Rubinstein-Avila et al., 2015; Téllez & Mosqueda, 2015; Wong-Fillmore, 2007).

Our results are also encouraging because our findings are consistent with current research about how often long-held beliefs by teachers may be challenging to change; however PSTs' experiences, including participation in teacher preparation programs or professional development, can contribute to developing strong beliefs about math instruction (Beach, 1994; Borg, 2011). Our work provides insight into teacher beliefs about math and language instruction in, therefore adding to the belief literature to provide evidence that teachers hold beliefs about effective language practices in bilingual mathematics contexts.

Mead's (1934) theories of belief formation and development shape the way we view the effects of the MALLI PD model. We point specifically to three social processes of the PST intervention that we suggest resulted in the PST' growth in knowledge and efficacy. First, the MALLI PD incorporated six anchor lessons into the mathematics methods course taught in Spanish. These lessons introduced and reinforced the MALLI language and literacy practices, and also modeled a translanguaging context for teachers to think about their own implementation of these strategies in their classrooms. Second, the CT mentors also played a major role in reinforcing the implementation of the MALLI practices, and CT provided critical feedback to help PSTs improve their integration of these strategies. Third, the lesson study peer analysis of recorded lessons, which integrated video club strategies to enhance collaboration and discussion around mathematics and language interaction. The PST recorded their instructional sessions, observed their own lessons and those of their peers, and engaged in semi-structured discussions with their peers. These lesson study meetings allow teachers to exchange insights derived from these dialogues within a smaller team as well as sharing their reflections with the entire group. Taken together, we found that the robust reinforcement of language and literacy practices in a translanguaging mathematics context contributed to the significant findings reported in this chapter. Our findings provide evidence for a particular PD model that can support the development of PSTs beliefs about integrating literacy practices in mathematics contexts. This model can be utilized within mathematics methods courses to provide PSTs opportunities to actively engage in literacy and math integration within their student teaching classrooms. These experiences can ultimately contribute to PSTs' beliefs and, in theory, shape their practices as they prepare to enter the teaching profession.

Limitations

Interpretations of the study results must consider several limitations. First, not all of the MALLI practices were implemented with equal efficacy, and this may be due to PSTs gravitating toward practices that resonated with their mathematical teaching dispositions, so they did not implement all the practices in a uniform fashion. Second, the sample size was small due to the time-intensive nature of classroom attrition. Therefore, these results must be interpreted carefully given the potential for selectivity bias related to participant attrition. However, our analysis included 40 PST, which is a noteworthy sample size for this type of study. Third, the participant sample was not selected at random, so our inferences are not causal, and our findings are not generalizable to populations of teachers outside of the teacher education programs we studied. Finally, we had no comparison group in this study, and thus no way to know if PST growth in efficacy was owing to maturation effects.

Future research will examine the extent to which improved changes in PSTs' beliefs about the role of literacy and biliteracy in mathematics classrooms influence changes in their practice. Thus, future analysis will focus on teacher's expressed belief about the role of literacy in mathematics, and the extent to which this belief is consistent with their integration of literacy practices in the classroom. As part of the MALLI project, we have collected classroom observation data that captured the extent to which PSTs integrated literacy practices in their teaching, so we are preparing to examine how practices relate to teachers' beliefs.

CONCLUSION

This study provides an important contribution to the literature on teacher beliefs and practice because it unravels elements of PD that can have an influence on PSTs in DLP. The results from this study show that the MALLI PD intervention had a significant effect on teachers' beliefs about the efficacy of integrating language and literacy practices in mathematics. Specifically, we found that critical elements that included: (a) the incorporation of anchor lesson in Spanish in the mathematics methods course, (b) providing a mentor teacher with training on the integration of language and literacy practices in math to provide guidance and feedback, (c) the lesson study collaborations that included video club activities, and (d) involving parents in math activities in two languages. Our results also point to the effectiveness of having a clear stance in terms of the role of language and the critical nature of availing a translanguaging context, where beyond using two languages, the goal is for EB students to have access to their full linguistic repertoires across languages to develop content knowledge, understanding, and reasoning as they develop English proficiency. This significant shift in PST beliefs provides a promising model for DLP to also potentially influence changes in their literacy and biliteracy practices in ways that maximize students' reasoning and conceptual understanding in mathematics.

REFERENCES

Avalos, M. A., & Secada W. G. (2019). Linguistically responsive teaching to foster ELL engagement, reasoning, and participation in a mathematics discourse community. In L. de Oliveira, K. Obenchain, R. Kenney, & A. Oliveira (Eds.), *Teaching the content areas to English language learners in secondary schools* (pp. 165–179). Springer.

Bahr, D. L., & Monroe, E. E. (2008). *An exploration of the effects of a practicum-based mathematics course on the beliefs of elementary preservice teachers.* https://www.cimt.org.uk/journal/bahrmonroe.pdf

Bandura, A. (1993). Perceived self-efficacy in cognitive development and functioning. *Educational Psychologist, 28*(2), 117–148.

Beach, S. A. (1994). Teacher's theories and classroom practice: beliefs, knowledge, or context? *Reading Psychology, 15*(3), 189–96.

Beck, I., McKeown, M., & Kucan, L. (2002). *Bringing words to life.* Guilford Press.

Biesta, J., & Tröhler, G. (2016). Introduction: George Herbert Mead and the development of a social conception of education. In J. Biesta & G. Tröhler (Eds.), *Philosophy of education* (pp. 1–16). Taylor & Francis.

Borg, S. (2011). The impact of in-service teacher education on language teachers' beliefs. *System, 39*, 370–380.

Bravo, M. A., Mosqueda, E., Solis, J. L., & Maldonado, S. I. (2022). Preparing teachers for dual language contexts: Strategies for teaching and assessing mathematics in two languages. In M. Machado-Casas, S. I. Maldonado, & B. Bustos Flores (Eds.), *Evaluating bilingual education programs: Assessing students' bilingualism, biliteracy, and sociocultural competence* (pp. 167–187). Peter Lang.

Celedón-Pattichis, S., Musanti, S., & Marshall, M. (2010). Bilingual teachers' reflections on students' native language and culture to teach mathematics. In M. Foote (Ed.), *Mathematics teaching and learning in K–12: Equity and professional development* (pp. 7–24). Palgrave Mcmillan.

Enderle, P., Dentzau, M., Roseler, K., Sutherland, S., Granger, E., Hughes, R., Golden, B., & Saka, Y. (2014). Examining the influence of RETs on science teacher beliefs and practice. *Science Teacher Education, 98*(6), 1077–1108.

Hansson, S. O., & Wassermann, R. (2002). Local change. *Studio Logica, 70,* 49–76.

Hume, D. (2007). *Enquiry concerning human understanding.* Hackett Publishing Company.

Johnson, K. E. (1994). The emerging beliefs and instructional practices of preservice English as a second language teachers. *Teaching and Teacher Education, 10*(4), 439–452.

Kajander, A. (2007). Unpacking mathematics for teaching: A study of preservice elementary teachers' evolving mathematical understandings and beliefs. *Journal of Teaching and Learning, 5*(1). https://doi.org/10.22329/JTL.V5I1.127

Knudsen, J., Stevens, H. S., Lara- Meloy, T., Kim H. J., & Shechtman, N. (2018). *Mathematical argumentation in middle school: The what, why, and how.* Corwin Mathematics.

Levitt, K. E. (2001). An analysis of elementary teachers' beliefs regarding the teaching and learning of science. *Science Education, 86*(1), 1–22.

Loeb, L. E. (1990). The priority of reason in Descartes. *The Philosophical Review, 99*(1), 3–43. https://doi.org/10.2307/2185202

Maldonado, S. I., Mosqueda, E., Bravo, M. A., & Solís, J. L. (2020). Assessing and teaching students' biliteracy in mathematics: A professional development model. *The Multilingual Educator,* 36–39.

Mead, G. H. (1934). *Mind, self, and society.* The University of Chicago Press.

Miller, D. L. (1973). George Herbert Mead: Symbolic interaction and social change. *The Psychological Record, 23,* 294–304.

Moschkovich, J. N. (2015). Academic literacy in mathematics for English learners. *Journal of Mathematical Behavior, 40,* 43–62. https://doi.org/10.1016/j.jmathb.2015.01.005

Mosqueda, E., Bravo, M., Solís, J. A., & Maldonado, S. I. (2022). Assessing emergent bilingual learners' mathematical biliteracy: Authentic mathematics writing assessment system. In M. Machado-Casas, S. I. Maldonado, & B. Bustos Flores (Eds.), *Assessment and evaluation in bilingual education* (pp. 223–244). Peter Lang.

Pajares, M. F. (1992). Teachers' beliefs and educational research: Cleaning up a messy construct. *Review of Educational Research, 62*(3), 307–332.

Polat, N., Mahalingappa, L., Hughes, E., & Karayigit, C. (2019). Change in preservice teacher beliefs about inclusion, responsibility, and culturally responsive pedagogy for English learners. *International Multilingual Research Journal, 13*(4), 222–238.

Rubinstein-Avila, E. B., Sox, A. A., Kaplan, S., & Mcgraw, R. H. (2015). Does biliteracy + mathematical discourse = binumerate development? Language use in a middle school dual-language mathematics classroom. *Urban Education, 50*(8), 899–937. https://doi.org/10.1177/0042085914536997

Rumsey, C., & Langrall, C. W. (2016). Promoting mathematical argumentation: These evidence-based instructional strategies can lead to deeper mathematical conversations in upper elementary school classrooms. *Teaching Children Mathematics, 22*(7), 413–419.

Savasci, F., & Berlin, D. F. (2012). Science teacher beliefs and classroom practices related to constructivism in different school settings. *Journal of Science Teacher Education, 23*(1), 65–86.

Schwitzgebel, E. (2002). A phenomenal, dispositional account of beliefs. *Nous, 36*(2), 249–275.

Téllez, K., & Mosqueda, E. (2015). Developing teachers' knowledge and skills at the intersection of language learners and language assessment. *Review of Research in Education, 39*(1), 87–131. https://doi.org/10.3102/0091732X14554552

Turner, E., Celedón-Pattichis, S., Marshall, M., & Tennison, A. (2009). "Fíjense amorcitos, les voy a contar una historia": The power of story to support solving and discussing mathematical problems with Latino/a kindergarten students. In D. Y. White & J. S. Spitzer (Eds.), *Mathematics for every student: Responding to diversity, Grades Pre-K–5* (pp. 23–41). National Council of Teachers of Mathematics.

Wong-Fillmore, L. (2007). English learners and mathematics learning: Language issues to consider. In A. H. Schoenfeld (Ed.), *Assessing mathematical proficiency* (pp. 333–344). Cambridge University Press.

CHAPTER 6

THE EFFECTS OF A NOVEL TEACHER PROFESSIONAL DEVELOPMENT MODEL ON STUDENT ACHIEVEMENT IN SPANISH/ENGLISH DUAL LANGUAGE PROGRAMS

Yuzhu Xia
Boston Public Schools

Liana Balloffet
University of California, Santa Cruz

Kip Téllez
University of California, Santa Cruz

Mathematics Instruction in Dual Language Classrooms, pages 97–114

www.infoagepub.com

ABSTRACT

The past decades witnessed an increasing popularity of dual language programs (DLP). However, issues such as the effective evaluation of DLPs and teacher professional development (TPD) for DLPs started to emerge (Xia et al., 2022). Part of this conversation involves the known debate on how TPD translates into better student outcomes in terms of assessment scores, and specifically, how difficult it is to measure the effects of TPD on student performance (e.g., Wallace, 2009; Yoon et al., 2007). This study takes all of the above into consideration, and contributes to the limited literature of the effectiveness of TPDs in relation to student performance. Specifically, this study examines how the Lesson Study With Video Clubs (LSVC) TPD model is effective in terms of students' literacy and mathematics performance (Xia et al., 2022).

For decades educators and educational researchers have been facing the challenge of how teacher professional development (TPD) translates into improved academic achievement (e.g., Garet et al., 2008; Kutaka et al., 2017). Despite the difficulty of identifying causal relationships and/or associations between TPDs and student outcomes in terms of standardized test scores, obtaining large-scale student outcome data associated with teachers who participate in TPDs has also been a great obstacle. Effective TPDs share some common features (e.g., Darling-Hammond & McLaughlin, 1995), but limited formats of TPDs have been successful so far due to various reasons such as relevance to content, sustainability, and accessibility, just to name a few. This study examines student data from two school districts, comparing student outcomes based on whether or not their teachers participated in a novel TPD, Lesson Study With Video Club (LSVC) in dual language programs (DLP). We aim to investigate if the LSVC TPD model, which focused on integrating literacy and mathematics instructional practices, was effective or not in terms of student outcomes in both English language proficiency and mathematics standardized testing results.

LITERATURE REVIEW/THEORETICAL FRAMEWORKS

Teacher Growth and Student Achievement

Although it is hard to imagine now, the professions (e.g., law, medicine, teaching), and the professional knowledge they require, were established only around 500 years ago, and they took a very long time to grow into the complex disciplines we now take for granted (Macdonald, 1995). Teaching, certainly considered a contemporary profession—or at least a semi-profession (Etzioni, 1969)—is a recent addition to the list, but teaching is not generally considered a complex profession when compared to medicine, for example. But a closer look at what defines a professional suggests that teaching shares most, but not all (e.g., salaries considerably higher than non-professional fields), of the features we associate with a profession. Like other professionals, teachers hold advanced, complex, esoteric, and arcane knowledge available only to those who have reached an advanced level (Macdonald, 1995). Another feature of a professional's knowledge is that it must grow and change with new advances, conditions, and technologies. The manner in which teachers gain new professional, pedagogical knowledge is the topic of our study.

In the field of medicine, for instance, physicians are required to update their knowledge of new methods and treatments and engage in approximately 150 hours every 3 years (states may differ) of continuing education in courses or programs approved by the state board of medicine. These requirements have been in place for decades. In contrast, teacher professional development (TPD) requirements are a more recent invention. In fact, the exact date is difficult to determine, but at some point in the 1980s, the educational community in the United States and elsewhere decided that teacher professional development was the most efficient strategy for raising student achievement. Perhaps the release of the scathing report *A Nation at Risk* (National Commission on Excellence in Education, 1983), at least in the United States, promoted the idea that teacher quality and teacher knowledge were crucial to improving student achievement. Exactly what kind of teacher knowledge or qualities result in increased student achievement remains contested (e.g., Harris & Sass, 2011).

In the intervening decades, we have witnessed the growth of a vast industry in TPD, as well as a range of academic papers suggesting what makes effective TPD (Desimone, 2009), even as the relationship between TPD and student achievement is cloudy (Garet et al., 2008). Nevertheless, the logic is nearly self-evident and can be distilled in this way: If educators are to increase student achievement, there are but two general strategies (Téllez & Mosqueda, 2015). To increase student learning educators can either (a) increase the time that teachers or other educators spend with students, or (b) make the time students spend with teachers or other educators more efficient. The relationship between time, efficiency, and learning can be expressed in the following equation: $E/T = L$, where E is efficiency in teaching, T is time, and L is learning. If you lack efficiency, you must add time to learning. If you have a fixed amount of time, you need to be more efficient in order to gain the same amount of learning. Efficiency, then, is the primary goal of TPD; that is, helping teachers to teach "more" in the same amount of time.

EFFECTIVE TPDS

It is uncommon to talk about TPD as increasing efficiency in teaching. However, teachers who grope for effective strategies with no underlying theory about what works or those who must use a crude form of private trial and error will take much longer to teach the same amount of material as a teacher who knows the most efficient methods. But if efficiency is not generally the recognized goal of TPD, what does the literature define as effective TPD? The research seems to converge on several features. For example, Darling-Hammond and McLaughlin (1995), and Borko (2004) both listed a set of features that are still largely true today (see Patthoff, this volume). More recently, Bigsby and Firestone (2017) identified, more or less, the same features. A focus on favorable student outcomes, along those lines, should always be considered as one important feature of effective TPDs; despite ongoing controversy on the matter (see Evans, 2022, for a review), student outcomes remain one, if not the best, quantifiable and comparable indicator of TPD's success. We argue that each of these features of effective TPDs might lead to teachers who teach more efficiently. In particular, the attention to teachers sharing effective practices within a particular teaching context (age, content, curricular goals) seems the most promising of strategies.

The Lesson Study With Video Club Teacher Professional Development

As a working group of educators and researchers on a grant-funded project designed to enhance elementary teachers' use of language in mathematics instruction in DLP (see Xia et al., 2022 for a description), we set out to devise a TPD that built on the features described above while also using emerging technologies (e.g., easy video recording, editing, and storage) to enhance mathematics education. The growing constraints on school systems seeking to provide quality TPD (e.g., lack of substitute teachers, growing costs, teachers' desire to spend as much time in the classroom as possible) also drove our use of the model. However, for a variety of reasons, which we will explore in the conclusion of our chapter, we sought to test our model omitting the school-wide feature of effective TPD. We called our new model LSVC (Xia et al., 2022). LSVC combines lesson study (Fernandez & Yoshida, 2004; Takahashi & Yoshida, 2004) and video club (Thompson, 2008; van Es, 2012; van Es & Sherin, 2008) to overcome the limitations of each model, and enrich them. The program we implemented coupled a loose structure with unified, relevant content, while focusing on integrating language, literacy, and mathematics instruction in DLP contexts (see Patthoff, this volume, for more details). Teachers also mentored a PST on how to integrate language into mathematics instruction in teaching mathematics using students' first language, as well as using formative assessment to assess student learning.

We began our TPD with 2½ days of in-person PD and one half-day virtual workshop, but shifted to a series of entirely virtual meetings during the COVID-19 pandemic. In both modalities, participants spent roughly 25–30 hours on LSVC TPD-related activities over the course of the academic year. This included about 12 hours of synchronous, full-group meetings, with the rest of participants' time spent working with teacher partners, administering and grading mathematical writing assessments, or planning and recording lessons. We observed that teachers found using LSVC effective because it is collaborative, affordable, accessible, and grounded in relevant content. Over the five years of the grant-funded project, over 60 teachers participated, endorsing the LSVC model with an enthusiasm we did not expect. Our study of the teachers' satisfaction with the model is documented in Xia et al. (2022), but we share a few of the positive comments here. For example, one teacher reported:

> One of the things that I found most helpful about the program was the amazing ideas from other teachers and particularly being able to watch examples of other teachers teach and focus on the different areas that were presented and see what it looked like in the classroom, as well as just seeing my teaching through the lens of others and seeing what kinds of things that they noticed and being able to see others teaching and recognize how they're teaching math. It really opened my eyes to how things might be done differently in different districts but still those basic needs of the students are still there. I was able to learn a lot of skills and strategies from those conversations with other participants of the program.

And another:

> We exchanged ideas, and it was just wonderful to see their work in action in those videos. And then, after that, the discussions that we had definitely deepened our understanding of how to teach better and how to think about different approaches, focusing on things that will make an impact on students' understanding in the area. So yeah, it was really helpful to have those interactions with other teachers and with our whole team ... the exchange of ideas was really fruitful.

Based on the teachers' endorsement of LSVC, we wondered if their positive reaction to LSVC (and the learning they gained) resulted in changes in their students' academic performance. We were encouraged by the research conducted by Kutaka et al. (2017) and Polly et al. (2015), for example, in spite of the fact that the TPDs in their study were much longer in duration than our project. The purpose of this chapter is to document the effects of LSVC on student achievement in mathematics and a test of English proficiency.

Specifically, we analyzed student outcomes for two of the school districts that we worked with, and focused on student gains for those who were in the classrooms of teachers who participated in the LSVC TPD.

Our formal hypotheses are as follows:

(1): *The LSVC TPD was effective in improving students' English language proficiency scores, and*

(2): *LSVC TPD was effective in students' mathematics achievement scores.*

METHODS

Data

We obtained student and teacher data from two school districts in California, one (School District A) is a traditional public school district and the other (School District B) is a charter school district. Table 6.1 shows the demographics information for both school districts. Both district A and district B have student demographics information and ELPAC and SBAC scores starting from the 2017–2018 school year (planning year of the TPD project) to the 2021–2022 school year. For SBAC data, neither district has outcome data for the pandemic years (2019–2020 & 2020–2021). District B also does not have ELPAC student outcome data for 2019–2020 and 2020–2021 due to the pandemic. The free/reduced lunch figures, however, are likely undercounts, owing to the COVID-19 era policy of providing free lunch to any student, whether or not their family completed the eligibility forms (California Department of Education, 2023).

Variables

The student outcome variables that are being used in this study include math assessment scores (CAASPP/SBAC), and language and literacy assessment scores (ELPAC).

TABLE 6.1 Demographics for Both School Districts

	Type	*N*	Hispanic	Spanish as First Language	English Learners (ELs)	Free/Reduced Lunch
District A	Public	7,253	79.44%	55.43%	42.08%	79.48%
District B	Charter	2,010	96.72%	86.82%	73.33%	69.35%

CAASPP/SBAC

The Smarter Balanced Assessment Consortium (SBAC) produces standardized tests aligned with the Common Core State Standards. These tests are adaptive online exams and are utilized in seventeen states. In California, the Smarter Balanced Summative Assessments for English language arts/literacy and mathematics are administered annually for students in Grades 3–8 and 11 as part of the California Assessment of Student Performance and Progress (CAASPP). Scores are reported both as achievement levels (1–4, "standard not met" to "standard exceeded"), and numerically as scaled scores. Scaled scores are measured on a continuous scale that rises as grade levels increase (ranging from approximately 2,000 to 3,000; California Department of Education [CDE], 2022). These scores provide insight into a student's current level of achievement and progress over time. When aggregated across a student population, scaled scores can depict changes in performance at the school and district level and highlight disparities in achievement between various student groups (The Regents of the University of California, 2022).

ELPAC

The English Language Proficiency Assessments for California (ELPAC) is an assessment designed to evaluate the English language proficiency of K–12 students in California whose primary language is not English. The ELPAC is administered online and consists of two parts: the initial assessment and the summative assessment. The initial assessment is administered to new students who have recently enrolled in a California school district and assesses their English language proficiency levels in listening, speaking, reading, and writing. The summative assessment is annually administered to students who have been identified as English language learners to measure their progress in English language proficiency throughout the school year in these same domains. The ELPAC is aligned with the English language development (ELD) standards in California and is used to monitor student progress and inform instructional decisions. The results of the assessment are used to determine if students require English language support services and to evaluate the effectiveness of these services.

This analysis uses summative ELPAC data. There are seven different tests that correspond to different grade levels: Kindergarten, Grade 1, Grade 2, Grades 3–5, Grades 6–8, Grades 9–10, and Grades 11–12. ELPAC scale scores are released as an overall score, as well as oral language and written language subscores. In this study, we were only able to obtain cumulative scale scores. Scale scores range from approximately 1,000 to 2,000. ELPAC scores in listening, reading, writing, and speaking are also reported with performance levels 1–4, where the students' skills in each domain are rated as "minimally developed," "somewhat developed," "moderately developed," or "well developed." Students who score within the 4 range on all four domains can be reclassified as fluent English proficient (RFEP), while students who have not reached level 4 will remain classified as EL and continue to receive ELD services (CDE, 2023). Unlike the SBAC, ELPAC scores are not on one continuous scale across grades. Each grade band test has its own scale score ranges that correspond to the different proficiency levels, which can be found online.[1] Examples of summative ELPAC practice tests can be found on the ELPAC website,[2] as well as test blueprints[3] and scale score ranges[4] for each grade span.

Predictor Variables

In this study, we included the following student demographics data as predictor variables given the data shared with us: (a) grade level, (b) gender, (c) race, (d) school, (e) IEP (individual educational plan) status/special education status, (f) primary language, and (g) English language learner status. We were unable to obtain other demographic information such as students' socioeconomic status information.

Participants

Over the course of 5 years during which the LSVC TPDs were conducted, a total of 18 teachers across the two districts were recruited in this current study, with years of experience ranging from 1 to 18 years (mean = 4.94, *SD* = 5.08). In District A, we have the student data for 1 participating teacher in 2017–2018; for 4 participating teachers in 2018–2019; 3 in 2019–2020; 4 in 2020–2021; and for 2 participating teachers in 2021–2022. In District B, we have the student data for 8 participating teachers in 2017–2018; for 10 participating teachers in 2018–2019; and for 2 participating teachers in 2021–2022.

DATA ANALYSIS

Out of the 7,253 students from District A, only 151 students have test scores for all ELPAC and SBAC assessments across the school years from 2017 to 2021, the period when LSVC was implemented. Out of the 2,010 students from District B, no students completed all ELPAC and SBAC assessments across the school years from 2017 to 2021. Therefore, multiple imputation was used to handle the missing assessment data for those who were missing one or more assessment data points.

Descriptives of student demographics were computed in *R* to compare student composition between the two school districts. Two-sample t-tests were performed in *R* to measure the difference in student ELPAC and SBAC scores between students who had a teacher who participated in LSVC TPD and those who did not. Multivariate analysis of variance (MANOVA) was performed to measure the effectiveness of the LSVC TPD on student outcome. The general MANOVA model is defined as below:

$$Y = SB + E$$

where Y represents the students' ELPAC or SBAC scores for the different school years, S is a vector of coefficients for the predictor variables, B is a vector of predictor variables, including but not limited to student demographics and if their teachers participated in the LSVC TPD, and E is a vector of random errors. Besides the student demographics information specified in previous sections, we added if the student's teacher participated in the LSVC TPD as an important predictor variable.

RESULTS

Descriptives

TABLE 6.2 Overall Special Education

	Yes	No
District A	13.26%	86.74%
District B	14.23%	85.77%

TABLE 6.3 Overall Race/Ethnicity

	American Indian or Alaska Native	Asian	Black or African American	Hispanic	Native Hawaiian or Other Pacific Islander	Two or More Races	White
District A	0.22%	15.43%	0.95%	79.44%	0.51%	1.74%	1.71%
District B	0.45%	0.40%	0.45%	96.72%	0.05%	0.25%	1.14%

TABLE 6.4 English Proficiency

	English Learner	English Only	Initially Fluent English Proficient (I-FEP)	Redesignated Fluent English Proficient	To Be Determined
District A	42.08%	33.53%	5.78%	18.57%	0.04%
District B	73.33%	12.38%	1.69%	12.59%	N/A

TABLE 6.5 Top 3 Primary Languages

	Spanish	English	Vietnamese
District A	55.43%	33.53%	6.96%
District B	86.82%	12.73%	0.20%

T-tests

Statistically significant differences between students with a teacher who participated in LSVC TPD and those whose teachers did not participate were found for multiple years of ELPAC scores in both districts. In District A, students of teachers who participated in LSVC TPD were found to have outperformed their counterparts on the ELPAC in school years 2019–2020, 2020–2021, and 2021–2022. In District B, LSVC teachers' students outperformed their counterparts on the ELPAC in years 2018–2019 and 2021–2022 (with 2019–2020 and 2020–2021 missing due to the pandemic). It is worth considering that this pattern did not appear in school year 2017–2018, the year prior to the PD implementation, for either district. We were also able to find statistically significant differences in students' SBAC scores between students of LSVC in 2021–2022 for District A. It is interesting to note that in District B students of LSVC teachers had significantly *lower* SBAC scores than their counterparts in school year 2017–2018, but did not have lower scores in the following two years of data collection during which the PD was being implemented. Tables 6.6 and 6.7 report all significant differences between students of participating and non-participating teachers for both districts.

Regression

Multivariate analysis of variance (MANOVA) was carried out in order to measure the effects of the MALLI PD on students' scores on the ELPAC and SBAC. For this portion of analysis it was necessary to use the imputed dataset, as the original dataset had missing assessment score values in one or

TABLE 6.6 District A I-Test Results

	t-value	*df*	*p*-value	95% CI
SY 2019–2020 ELPAC scale score	9.6192	46.863	1.138e-12	[55.00512, 76.45480]
SY 2020–2021 ELPAC scale score	2.8898	127.58	0.03891	[0.9912583, 37.1390183]
SY 2021–2022 ELPAC scale score	4.1662	140.04	0.001488	[10.49726, 42.98968]
SY 2021–2022 SBAC scale score	2.5167	137.32	0.013	[5.941011, 49.510663]

TABLE 6.7 District B *T*-Test Results

	t-value	*df*	*p*-value	95% CI
SY 2017–2018 math SBAC scale score	6.5419	136.03	1.136e-9	[47.06102, 87.84088]
SY 2018–2019 ELPAC scale score	3.0084	241.62	0.002904	[6.009502, 28.805958]
SY 2021–2022 ELPAC scale score	3.7606	125.49	0.0002588	[10.98765, 35.39914]

more tests to perform MANOVA. For District A, outcome measures included the ELPAC scores for school years 2017–2018 through 2021–2022, and math SBAC scores for school years 2017–2018, 2018–2019, and 2021–2022 (2019–2020 & 2020–2021 school years data missing due to the pandemic). For District B, outcome measures consisted of ELPAC and math SBAC scores for school years 2017–2018, 2018–2019, and 2021–2022 (2019–2020 & 2020–2021 school years data missing due to the pandemic).

In District A all indicators were found to be statistically significant, while in District B student race and gender were not found to be significant indicators. It is possible that the lack of significance for student race in District B is due to the homogeneity of the district's student body (96.72% of students reported Hispanic). In both districts, indicators were significant at the $p < .001$ level.

TABLE 6.8 MANOVA Model Output for District A

	Df	Pillai	approx F	num Df	den Df	Pr (>F)
LSVC_Teacher_Y	1	0.005653	5.143	8	7237	2.046e-06 ***
Spec_Ed	1	0.104841	105.950	8	7237	< 2.2e-16 ***
Current_Grade_Level	1	0.210669	241.484	8	7237	< 2.2e-16 ***
School_Name	1	0.003714	3.372	8	7237	0.0007239***
Reported_Race	1	0.036653	34.419	8	7237	< 2.2e-16 ***
Gender	1	0.006772	6.168	8	7237	5.838e-08 ***
English_Proficiency	1	0.157512	169.130	8	7237	< 2.2e-16 ***
Student_Primary_Language	1	0.018851	17.381	8	7237	< 2.2e-16 ***
Residuals	7253					
Signif. Codes: 0 '***' 0.001 '**' 0.01 '*' .05 '.' 0.1 ' ' 1						

TABLE 6.9 MANOVA Model Output for District B

	Df	Pillai	approx F	num Df	den Df	Pr (>F)
LSVC_Teacher	1	0.038325	13.258	6	1996	8.902e-15***
IEP_Status	1	0.009425	3.165	6	1996	0.0043014**
Current_Grade_Level	1	0.292950	137.833	6	1996	<2.2e-16***
School_Name	1	0.078543	28.356	6	1996	<2.2e-16***
Race	1	0.004347	1.452	6	1996	0.1303641
Gender	1	0.000640	0.213	6	1996	0.9728454
English_Language_Learner_Status	1	0.072784	26.114	6	1996	<2.2e-16***
Primary_Language	1	0.012258	4.128	6	1996	0.0003975***
Residuals	2001					

Signif. Codes: 0 '***' 0.001 '**' 0.01 '*' .05 '.' 0.1 ' ' 1

DISCUSSION

We examined student achievement in ELPAC and SBAC scores across the years when the LSVC PD was in preparation and implemented, and showed that students with teachers who participated in the LSVC PD presented consistent improvement in their ELPAC scores and one year of SBAC scores for District A. The fact that their teachers participated in the LSVC PD was statistically significant in the model indicating that the LSVC PD format was beneficial to the students in terms of English proficiency and mathematics development.

Students of teachers who participated in LSVC TPD showed consistent increase in ELPAC and SBAC scores.

Despite issues with incomplete data, our analysis revealed a clear pattern—students' ELPAC scores were consistently higher when they had been taught by a LSVC teacher. This pattern persisted in all years in which LSVC PD was implemented, and for which we had data. For SBAC scores, statistically significant results were found in district A for the school year of 2021–2022. It is especially noteworthy that several of the participating teachers were teaching mathematics in Spanish. We must also reiterate that the LSVC TPD was able to achieve these results without a strong focus on direct instruction of teachers, but rather with a relaxed approach in which the PD sessions provided a vehicle for facilitating teachers' exchange of ideas as peers and mentors for one another. It appears that through LSVC TPD, regardless of the teacher's language of instruction, a pedagogical focus on integrating discourse, literacy, and vocabulary into mathematics instruction was able to accelerate students' acquisition of English and mathematics.

This corroborates existing literature in documenting essential features of effective TPDs and how teachers learn best (e.g., Borko, 2004; Darling-Hammond & McLaughlin, 1995). What separates the LSVC TPD and traditional TPD formats is the workshops provided for teachers teaching in the same specialized settings with one focused content/area, giving the teachers themselves an opportunity to play an active and the central role in their own learning and form their own professional community (Lave & Wenger, 1991). Specifically, the focus on the integration of language and literacy in their mathematics instruction in both Spanish and English provided theoretical background and hands-on approaches to enrich their mathematics instruction through better usage of the two languages and their funds of knowledge.

Additionally, the LSVC TPD model caters to teachers with various levels of experience (Xia et al., 2022). Regardless of teachers' years of experience whose data was used for this chapter, students with teachers who participated in the LSVC TPD workshops showed consistent improvement in their ELPAC scores after the LSVC TPD was implemented. The fact that their teachers participated in the LSVC TPD was statistically significant for both school districts regardless of their experience level indicates that the LSVC TPD is effective and beneficial for teachers with various backgrounds teaching in specialized settings, such as the DLPs (Xia et al., 2023).

The Issue of Missing Data and the Challenge of Obtaining Student Data

Although we were able to find statistically significant differences in both literacy ELPAC and math SBAC scores with the data available, the raw data contained a large amount of missing data, especially for SBAC results. Not only were all scores from 2019–2020 and 2021–2021 missing due to testing disruptions caused by the pandemic, but for the years we had test data available, there were also more missing SBAC scores than ELPAC scores. Missing data decreases the power of our statistical tests and therefore weakens our analysis. This disparity can be attributed to the SBAC being administered only to students in Grades 3 and up, resulting in a lack of scores to compare for teachers working in lower elementary classrooms.

On top of the issue of incompleteness of the student data, we have experienced great challenges in obtaining student data from a total of 11 school districts we have worked with in California and Texas over the course of 5 years. Due to the school districts' capacity and other logistical constraints, it took the evaluation and research team years to receive the two school districts' data, which still has a relatively large proportion of missing data. This again speaks to the shared dilemma the educational research field has been experiencing in connecting effective TPDs with student achievement (e.g., Garet et al., 2008).

CONCLUSIONS AND IMPLICATIONS

Our results add to the growing body of research demonstrating that TPD can result in gains in academic achievement, particularly when TPD follows the recommendations of the past 4 decades. Specifically, our TPD was content-focused (i.e., mathematics instruction) which other studies of effective TPD have shown to a key component (Katuta et al., 2017). We also found that our LSVC model resulted in gains in English learning, which went beyond the content focus of mathematics. Similarly, Mutch-Jones, Hicks, and Sorge (2022) found that a TPD centered on science instruction yielded performance gains in mathematics and ELA as well as science. Our LSVC model built upon the research demonstrating the effectiveness of video as a tool for teacher growth (e.g., Patthoff, 2022; Sherin & Russ, 2014) as perceived by teachers, but we demonstrated that LSVC also resulted in gains in student achievement. Indeed, we would argue that video analysis of participating teachers' instruction, by the teachers themselves, should be added to the list of effective TPD characteristics.

While corroborating existing research, the TPD in our study did not adhere to a few of the common assumptions of effective TPD. For instance, we discovered student effects without engaging in school-wide TPD. Of course, we do not know if the results would have been more robust if the teachers in our study would have been part of a school-wide LSVC, but we suspect not. We have several reasons for our conclusion. First, to gain the support of school leaders is challenging because they will expect that any TPD is connected to other aspects of school change, with which teachers may or may not agree. We were able to get school administrators to "buy" into LSVC because we did not bring in a new curriculum or specific method. We emphasized that we were helping teachers understand the relationship between language and mathematics instruction more fully, which is a key feature of the Common Core State Standards. Thus, our participating teachers' administrators were willing to allow only interested teachers to take part. Second, school-wide TPD includes all the school's teachers, and some teachers do not respond well to the type of TPD LSVC demands. A deeply reflective TPD model such as LSVC requires dedicated teachers who are motivated to improve their practice at the right time in their careers. Third, like the specialties in the medical field (e.g., obstetrics, dermatology, oncology) classroom instruction has become highly specialized and any attempts at generic TPD will likely fail. Our study has shown that LSVC allowing motivated teachers, working in similar contexts, teaching students of similar ages, and addressing similar content, to work together on problems of practice, makes sense.

The other key finding is that no "expert" is needed for effective TPD. In our TPD, we introduced the teachers to some new ways of thinking about mathematics and language, and invited them to consider how literacy, discourse, and vocabulary might be developed within mathematics lessons, but beyond our initial 2-hour presentation and 15-minute introductory presentations at the beginning of group meetings, the teachers were on their own to design lessons, watch one another's lessons, and reflect together on the implications.

The LSVC TPD was mainly focused on the integration of mathematics, language, and literacy in DLP classrooms. Without direct and explicit "teaching" of how to achieve that (the research team presented the theoretical framework of this concept at the initiation of the year-long cycle), teachers somehow grasped how to work it out through conversations with peer teachers, recording, watching, and debriefing classroom videos, and extended discussions even after the project was completed. Regardless of the language of instruction, students with teachers who participated in the LSVC TPD showed statistically significant improvements in their ELPAC and SBAC scores in both school districts. The large amount of missing data calls for greater effort and collaboration among schools, researchers, and all other educators to work together to study effective TPDs for specialized programs like DLPs to achieve better student outcomes in terms of standardized testing.

The results of studies on the effect of DLP participation on student academic performance seem almost magical (e.g., Collier & Thomas, 2004). For instance, Alanís (2000) found that students who participated in a DLP for at least 3 years had the highest mathematics scores while students who had participated in the program for two years or less had the lowest. However, this result could easily be the result of program attrition (i.e., only the most academically capable students stayed in the program). Alanís recognized this potential but did not investigate it. And neither Alanís nor Collier and Thomas mention teacher professional development or background characteristics in their studies. In contrast to other inquiries concerning student achievement in the DLP context, our study compared student outcomes from two groups that both attended DLP, thus essentially controlling for the positive effects of this program model. Through this research model, we found that LSVC in a DLP context was not only perceived as effective for participating teachers, but also improved student performance. Given researchers' ongoing struggles to connect TPD with evidence of student achievement, especially for students from historically underserved communities, coupled with persistent barriers to the professionalization of the teaching field, we believe our findings are worthy of attention and perhaps suggest wider adoption of LVSC or other, similar models. Researchers and educators should consider contextualizing LVSC in other, if not all, intervention activities that may have led to positive student outcomes. Admittedly, besides the delivery model, teachers' instruction also had an impact on the observed student outcome improvements. The LSVC model made it easier for teachers to bring useful and approachable techniques into their mathematics instruction throughout the year and onwards. It was the combination of an effective TPD model and teachers' instruction that embraced what they gained from the TPD that made student outcome improvements possible and consistent. We hope that the promising evidence provided in this chapter encourages school districts, teachers, and communities to reconsider their TPD and implement models that respect the profession of teaching.

ACKNOWLEDGMENT

This research was supported by a grant from US Department of Education, Office of English Language Acquisition, National Professional Development Program under Grant # [T365Z170070].

1. We recognize that not all TPD is designed to raise student achievement. For instance, if a school or school system adopts a new attendance system that requires teachers to understand new technology and processes, such a TPD might be very useful but not necessarily expected to raise student achievement.
2. We also acknowledge the challenge educators and educational researchers are facing when student achievement data is needed to investigate the effectiveness of TPDs or any interventions in educational settings, due to uncontrollable factors such as the pandemic, constraints regarding sharing individual student information, and the missingness of standardized testing results.

NOTES

1. https://www.cde.ca.gov/ta/tg/ep/documents/summativescalescores.pdf
2. https://www.elpac.org/resources/practicetests/
3. https://www.cde.ca.gov/Ta/tg/ep/documents/elpacsummativebluprt.pdf
4. https://www.cde.ca.gov/ta/tg/ep/documents/summativescalescores.pdf

REFERENCES

Alanís, I. (2000). A Texas two-way bilingual program: Its effects on linguistic and academic achievement. *Bilingual Research Journal, 24*(3), 225–248.

Bigsby, J. B., & Firestone, W. A. (2017). Why teachers participate in professional development: Lessons from a schoolwide teacher study group. *The New Educator, 13*(1), 72–93.

Borko, H. (2004). Professional development and teacher learning: Mapping the terrain. *Educational researcher, 33*(8), 3–15.

California Department of Education. (2023). *EdData—Education data partnership.* Retrieved June 9, 2023 from http://www.ed-data.org/.

California Department of Education, Educational Testing Service. (2022). *Smarter balanced assessments for English language arts/literacy and mathematics.* https://www.caaspp.org/administration/about/smarter-balanced/

California Department of Education, Educational Testing Service. (2023). The ELPAC website. https://www.elpac.org/

Collier, V. P., & Thomas, W. P. (2004). The astounding effectiveness of dual language education for all. *NABE Journal of Research and Practice, 2*(1), 1–20.

Darling-Hammond, L., & McLaughlin, M. W. (1995). Policies that support professional development in an era of reform. *Phi Delta Kappan, 76*(8), 597–604.

Desimone, L. M. (2009). Improving impact studies of teachers' professional development: Toward better conceptualizations and measures. *Educational Researcher, 38*(3), 181–199.

Etzioni, A. (1969). *The semi-professions and their organization: Teachers, nurses and social workers.* Free Press.

Evans, L. (2022). Doubt, skepticism, and controversy in professional development scholarship: Advancing a critical research agenda. In I. Menter (Ed.), *The Palgrave handbook of teacher education research* (pp. 1–25). Palgrave Macmillan.

Fernandez, C., & Yoshida, M. (2004). *Lesson study: A Japanese approach to improving mathematics teaching and learning.* Routledge.

Garet, M. S., Cronen, S., Eaton, M., Kurki, A., Ludwig, M., Jones, W., Uekawa, K., Falk, A., Bloom, H. S., Doolittle, F., Zhu, P., & Sztejnberg, L. (2008). *The impact of two professional development interventions on early reading instruction and achievement.* NCEE 2008-4030. National Center for Education Evaluation and Regional Assistance.

Harris, D. N., & Sass, T. R. (2011). Teacher training, teacher quality and student achievement. *Journal of Public Economics, 95*(7), 798–812.

Kutaka, T. S., Smith, W. M., Albano, A. D., Edwards, C. P., Ren, L., Beattie, H. L., Lewis, W. J., Heaton, R. M., & Stroup, W. W. (2017). Connecting teacher professional development and student mathematics achievement: Mediating belonging with multimodal explorations in language, identity, and culture. *Journal of Teacher Education, 68*(2), 140–154.

Lave, J., & Wenger, E. (1991). *Situated learning: Legitimate peripheral participation.* Cambridge University Press.

Macdonald, K. M. (1995). *The sociology of the professions.* SAGE Publications.

Mutch-Jones, K., Hicks, J., & Sorge, B. (2022). Elementary science professional development to impact learning across the curriculum. *Teaching and Teacher Education, 112,* 103625.

National Commission on Excellence in Education. (1983). A nation at risk: The imperative for educational reform. *The Elementary School Journal, 84*(2), 113–130.

Patthoff, A. J. (2022). *Exploring pre-service teachers' learning of formative assessment in elementary, multilingual classrooms.* University of California.

Polly, D., McGee, J., Wang, C., Martin, C., Lambert, R., & Pugalee, D. K. (2015). Linking professional development, teacher outcomes, and student achievement: The case of a learner-centered mathematics program for elementary school teachers. *International Journal of Educational Research, 72,* 26–37.

Sherin, M. G., & Russ, R. S. (2014). Teacher noticing via video: The role of interpretive frames. In B. Calandra & P. Rich (Eds.), *Digital Video for Teacher Education* (pp. 11–28). Routledge.

Takahashi, A., & Yoshida, M. (2004). Ideas for establishing lesson-study communities. *Teaching Children Mathematics, 10*(9), 436–443.

Téllez, K., & Mosqueda, E. (2015). Developing teachers' knowledge and skills at the intersection of English language learners and language assessment. *Review of Research in Education, 39*(1), 87–121.

The Regents of the University of California. (2022). *Smarter balanced validity research.* https://validity.smarterbalanced.org/scoring/

Thompson, A. (2008). *Using video technology to provide a professional development forum for reflection on the use of academic language for mathematics in elementary school teachers* [Conference presentation]. The California Mathematics Council North, Asilomar, CA.

van Es, E. A. (2012). Examining the development of a teacher learning community: The case of a video club. *Teaching and Teacher Education, 28*(2), 182–192.

van Es, E. A., & Sherin, M. G. (2008). Mathematics teachers' "learning to notice" in the context of a video club. *Teaching and Teacher Education, 24*(2), 244–276.

Wallace, M. R. (2009). Making sense of the links: Professional development, teacher practices, and student achievement. *Teachers College Record, 111*(2), 573–596.

Xia, Y., Patthoff, A., Bravo, M., & Téllez, K. (2022). "We don't observe other teachers": Addressing professional development barriers through lesson study and video clubs. *Teacher Learning and Professional Development, 7*(2022).

Xia, Y., Patthoff, A., & Balloffet, L. (2023). Teacher growth across the career spectrum in lesson study with video club professional development model [Manuscript submitted for publication]. Education Department, University of California, Santa Cruz.

Yoon, K. S., Duncan, T., Lee, S. W.-Y., Scarloss, B., & Shapley, K. L. (2007). *Reviewing the evidence on how teacher professional development affects student achievement* (Issues & answers; REL 2007-No. 033). Regional Educational Laboratory Southwest (NJ1).

CHAPTER 7

TEACHER PROFESSIONAL LEARNING IN DUAL LANGUAGE PROGRAMS

Adria Patthoff
Ventura College

ABSTRACT

This chapter presents an overview of studies exploring the effective features of professional learning programs, with a particular focus on the intersection of dual language programs, mathematics instruction, and elementary settings. Beginning teachers experience challenges distinct from teachers with decades of experience, and therefore have different needs when it comes to supporting their teaching and professional development (Feiman-Nemser, 1983). Across the career spectrum, and within and among teachers' grade and subject specializations, school administrators must contend with how to differentiate professional learning programs to support their faculty's diverse learning needs. This chapter discusses features of contexts that enhance and promote teacher professional learning in dual language programs, highlighting mathematics instruction. Considerations include the range of professional experiences of classroom faculty, diversity of teachers' grade and subject

Mathematics Instruction in Dual Language Classrooms, pages 115–134

specializations, cost-effectiveness, and sustained engagement with professional learning programs and communities.

Across these chapters, learning is described as an iterative process of moving toward expertise (e.g., Kelly, 2006, p. 514). At its most effective, a professional teacher does not become an expert spontaneously, in isolation, or in one step. One way teachers learn and develop is when they observe their students and take time to discuss these observations with fellow teachers. They learn and develop when they are systemically and thoughtfully provided structured and semi-structured opportunities to critically explore practice-informing theories while receiving guidance from peers and experts, as they apply what they learn in their own classrooms (e.g., Borko, 2004; Darling-Hammond & McLaughlin, 1995). A commonly cited feature of effective teacher professional development (TPD) is frequent opportunities for teachers' mutual collaboration (Kizilbash, 2020), enacted through structures that are sustained, ongoing, intensive, and supported (Darling-Hammond & McLaughlin, 1995). In the specialized context of dual language programs, it is additionally critical for TPD content to attend to two overlapping categories of teachers' knowledge: pedagogical content knowledge (Shulman, 1987) and pedagogical language knowledge (Bunch, 2013; Galguera, 2011).

This chapter is divided into three parts. The first section briefly reviews research on effective features of teacher professional learning and development, with a specific focus on the needs of elementary teachers working in dual language programs teaching mathematics. The second section presents a detailed examination of a particular hybrid TPD structure—Lesson Study With Video Club. This structure illustrates one method of TPD that empirically supports teachers' professional growth in the dual language program (DLP) environment. The last section offers generalized recommendations for developing more effective and teacher centered TPD experiences in specialized contexts.

EFFECTIVE FEATURES OF TEACHER PROFESSIONAL LEARNING AND DEVELOPMENT

Before diving into specific features of professional development that support teacher learning, it is useful to distinguish between two common outcomes to workshops and programs aimed at developing teachers' expertise. Fullan and Hargreaves (2016) describe the outcomes of professional learning and professional development as a set of similar but distinct paths. *Professional learning* is concerned with content and process, or "what and how teachers learn"; *professional development* is more about reflection, community, and identity as professional, and "would include 'mindfulness' and 'team building' as more holistic aspects of the teacher learning process" (p. 63). This distinction is a useful tool for considering the research on effective features of TPD. Programs, workshops, seminars, and other formats meeting the criteria for *both* professional development and professional learning tend to be more successful in promoting and producing lasting effects on teachers' practice. Programs that isolate these concepts and goals, commonly seen as one-day in-service events that often occur prior to the school year or during professional development days, often fall short of meeting the combined requirements for both goals.

Darling-Hammond and McLaughlin's (1995) list of effective TPD features continues to be relevant. They recommended the following characteristics: (a) engage teachers in concrete tasks of teaching, assessment, observation, and reflection; (b) structure activities using participant-driven inquiry, reflection, and experimentation; (c) embrace collaboration; (d) connect to and derive from teachers' work with students; (e) be sustained, ongoing, intensive and supported by modeling, coaching, and the collective solving of specific problems of practice; and (f) connect to other aspects of school change (p. 598). These basic principles are still valid. However, since the 1990s, technology advances have increased opportunities to communicate and collaborate between schools. Schools are no longer limited due to geography or lack of access to consistent, timely, and effective communication tools. Before introducing a modern model of TPD that utilizes technology, two central themes from Darling-Hammond and McLaughlin's (1995) list merit further description: collaborative/interactive and sustained/ongoing.

Collaborative/Interactive TPD

In general, a commonly cited feature of strong TPD is that teachers must collaborate with each other to make genuine professional growth (Kizilbash, 2020). Relatedly, this collaboration and the content of the TPD should be clearly contextualized in participating teachers' own classrooms. Further, collaboration among teachers with heterogeneous backgrounds is a proven benefit to teachers' growth (e.g., Putnam & Borko, 2000). In diverse groups, TPD participants draw on each other's experiences and insights through meaningful discussion and reflection (Kizilbash, 2020). Diversity provides new perspectives. It also has the potential to decrease the levels of comparative affect and discomfort with vulnerability, often found in groups of teachers who have the same grade level, years of experience, utilize the same curriculum, etcetera. For mathematics teachers, semi-structured discussions and facilitated activities that center collaborative problem-solving are common aspects of professional development perceived effective by participants (Fernandez & Yoshida, 2004; van Es & Sherin, 2008). The constructs of collaboration and interaction meet the criteria for both professional development and professional learning (Fullan & Hargreaves, 2016). Through conversations and collaboration, teachers build rapport and community through mindful interactions. Teachers learn from listening to others' experiences and deepen their understanding of their own experiences through answering questions and responding to comments from peers.

Sustained/Ongoing TPD

The literature consistently shows strong, positive correlations between implementation of PD models sustained over time and student achievement (Lee et al., 2008; Llosa et al., 2016; Tong et al., 2017). TPDs completed over several days throughout a school year, with breaks in between for teachers to reflect, practice, and apply skills and concepts, are shown to be effective (Lee et al., 2008; Tong et al., 2017). Models are effective when they are (a) collaborative, offering ample opportunities for teachers to discuss and work together; (b) spread over time, enabling participants to practice methods discussed during sessions and come back to reflect together; and (c) provide explicit models. As above, this feature meets the criteria for both professional development and professional learning. Community is built over time and through repeated, meaningful opportunities to interact, with breaks between conversations. These breaks and pauses allow for reflection, application, and re-consideration of ideas presented during each TPD event. For example, Maldonado and colleagues (2021) conducted a series of three professional workshops with time between for mathematics teachers to implement and consider revisions to a formative assessment examining the intersection between language practices and mathematical proficiencies. The authors, like others (e.g., Xia et al., 2022), argued that breaks between the structured conversations, time to put to practice and consider the assessment tool beyond their own classroom, provides an environment for sustaining conversation; further, for enhancing teachers' learning.

Special Considerations for TPD in Dual Language Programs

Teachers and teaching are specialized, whether by grade level or content area, by programmatic and environmental structures, among many other possible features. TPD, as such, has the opportunity to attend to the specialized nature of teaching and cater to the specific needs of each teacher while simultaneously allowing for creativity and learning through collaborative interaction with others in one's own school and across the globe. Most relevant to this book: DLP teachers benefit from expanding their professional networks beyond their grade level teams and their singular, specialized school environment (Moloney & Wang, 2016). In comparison to the quantity of monolingual public schools in the United States, there are far fewer DLPs (American Councils Research Council, 2022), a significant obstacle in developing professional networks around this specialization.

The traditional method of "bringing in an expert" to present content to a passive group of teachers is outdated and often ineffective. The connective opportunities of social media, video messaging platforms, and other technical tools have enhanced our ability to communicate and share knowledge. The common and accepted TPD conceptions of "school-wide change," the geographical location of participants who collaborate, and the methods and places in which sustained interactions take place must be broadened.

Borko (2004), expanding on Darling-Hammond and McLaughlin's (1995) list, highlights the challenge of expanding or "scaling up" effective TPD. Borko (2004) argues that effective TPD has the following attributes: (a) focuses on the pedagogical content and methods that teachers use teach (Shulman, 1987); (b) provides active and practical opportunities for teachers to try out ideas and receive feedback, including examining student-made artifacts; (c) is centralized around a theme or issue and provides consistent opportunities for teachers to get feedback, practice and learn from trial-and-error in the classroom; and (d) is collective so that teachers have the opportunity to share practices with each other and to give and receive feedback (Marks & Louis, 1999). Borko's characteristics center on the importance of attending to teachers' current and specific contexts, calling attention to the integration of content and relevant student work. More recently, Bigsby and Firestone (2017) identified a similar set of features necessary for effective TPD.

While these feature lists are useful, they are all limited in that they view TPD as something that must happen in person and synchronously. My colleagues and I argue for an expanded conception of where, when, and how TPD takes place (Xia et al., 2022). Advocates for virtual professional development (VPD), including formats of online blogs and other media, propose that VPD has advantages over traditional formats. VPD is informal, individualized, and accessible, while at the same time maintain clear goals and structures (e.g., Irby et al., 2015; Lynch et al., 2021). VPD models, however, do not always explicitly invite or coordinate direct collaboration and conversation between participants. The next section presents a hybrid model of TPD, expanding on the structures of VPD and traditional TPD formats. This model includes the list of features of effective TPD, described above. It also utilizes features of VPD, including live, virtual and asynchronous activities. Further, it has statistically demonstrated significant potential to meet the needs of diverse teachers in a variety of specialized contexts, including dual language programs, across the career spectrum, from preschool through middle school, and those teaching in subject-specific settings.

LESSON STUDY WITH VIDEO CLUB: AN EXAMINATION OF ELEMENTARY MATHEMATICS DUAL LANGUAGE TEACHERS' EXPERIENCE IN A HYBRID TPD MODEL

The work of the Mathematics and Language, Literacy Integration (MALLI) in dual language settings project is one example of how TPD might be expanded beyond the list of school/district-specific change, learning, and development. The MALLI project attends to the specialized nature of teaching while taking advantage of technological advances to blur the geographic and curricular policy boundaries of dual language schools and districts. Combining two highly effective models of PD—lesson study and video club—the year-long format provides professional development to improve mathematical learning and teaching in DLPs. The project also meets the needs of teachers across the career spectrum, including those in teacher preparation programs, in their first years of teaching, and teachers who have more years of experience. This project involves preservice (student) teachers, mentor (cooperating) teachers, and parents, and through ongoing and overlapping activities. Returning to the themes above, the project fully embodies the effective features of collaboration/interaction as well as sustained/ongoing, all while extending beyond the traditional notion of TPD occurring on a single staff development day, isolated to one school or district.

The Hybrid Model: Lesson Study + Video Club

MALLI uses a hybrid TPD model to encourage reflection and collaboration through workshops and meetings that occur over the course of a school year. The model attends to challenges common to expanding and "scaling up" by using a variety of presentational and participatory mediums to reduce the need to travel and schedule substitutes. The design combines aspects of two TPD models proven to be effective in the literature: lesson study and video club, or LSVC.

The first component of LSVC is lesson study, a model originating in Japan (Fernandez & Yoshida, 2004; Takahashi & Yoshida, 2004). The essence of lesson study is *konaikenshu*, which means in-school (*konai*) training (*kenshu*; Fernandez & Yoshida, 2004). "In school training" differs from what we imagine in American schools, in that it brings together the entire teaching staff of a school for hours to collaborate on an agreed-on school goal and action plan. In a lesson study approach, teachers conduct cycles of inquiry to ultimately facilitate student learning. The TPD model MALLI uses commits to a year of structured activities, reflection and support via partnerships, large-group discussions, and instructional coaches. Two features of lesson study have been found to be very effective, collaboration among teachers and the teach-reteach strategy for improvement (e.g., Cajkler et al., 2014; Coenders & Verhoef, 2019). Despite the proven effectiveness of Lesson Study, the prototypical model has significant limitations. For example, teachers must be physically present with each other before, during, and after lesson enactment. The lesson study model could reach more individuals if it incorporated elements of video, recordings of teachers' lessons as well as virtual conferencing platforms.

The second component of the PD model MALLI uses is Video Club (Thompson, 2008; van Es & Sherin, 2008; van Es, 2012). Research consistently suggests that teachers who video recording their lessons and watch themselves in small groups supports their ability to reflect on and therefore improve instruction, subsequently leading to improved student learning (e.g., Borko et al., 2008; Brantlinger et al., 2011). The medium of video-recorded lesson enactments combined with structured, cyclical, rigorous discussions about recordings helps build a professional learning community (van Es, 2012; Alles et al., 2019).

Simply bringing teachers together does not guarantee building a professional learning community, in lesson study or in video clubs. Utilizing and enforcing co-created rules of discourse among participants is necessary for teachers to establish a learning atmosphere for productive conversations regarding classroom dialogue as seen in videos of participants' own classrooms (Alles et al., 2019). Outside facilitators are useful to help keep conversations focused, as participating teachers sometimes deviate from the specific focus of the club, and discuss more general topics (Alles et al., 2019). Despite the distinct advantages of the lesson study and video club models, there are limitations to both, when used separately. Most limiting is the necessity for teachers to be physically present, which discourages relationships across geographic boundaries and is difficult to coordinate amongst teachers' daily school schedules. This is critical, as teacher's instructional context (e.g., dual language program, newcomer student body) may be more similar to schools located at a distance from the teacher's own school.

The LSVC format used in the MALLI project accounts for both professional learning and professional development (Fullan & Hargreaves, 2016). First, the "what" and "how," professional learning, is visible through the specialized content discussed and practiced through video and semi-structured discussions. The format also functioned as professional development, with features included to help participants establish and maintain a teacher professional community, for example partnered discussions and regular interactions over a year-long period to build rapport. In this project, teachers are united by school context (dual language programs) and learning focus (integration of language and mathematics) but differ in grade taught and years' experience. They are provided consistent opportunities to collaborate on lesson plans and reflections and to communicate with each other about specific MALLI content they find valuable for their teaching practice.

We examined the perceived quality and influence of experiences of three cohorts of Spanish/English DLP teachers participating in live and online MALLI TPDs centered on specialized content over year-long periods for three years. Participating teachers came from 31 schools, representing 11 school districts and two states. Each year of the LSVC consists of three parts: (a) developing common understandings of concepts, skills, and practices; (b) applying concepts, skills, and practices in teachers' own classrooms; and (c) engaging in semi-structured partnered, small group, and large group feedback and reflection conversations. The graphic shown in Figure 7.1 illustrates the three parts and the timeline utilized for this version of LSVC.

Part I: Developing Common Understandings

The content of the TPD in this version of LSVC was designed to assist elementary DLP teachers to deliberately integrate one of three types of "pedagogies" foci (literacy, vocabulary, or discourse) into their mathematics instruction. In the introductory workshop, university-based facilitators discussed the integration of literacy, discourse, and vocabulary in DLP schools. This workshop is the only in the series to utilize a portion of its time in a

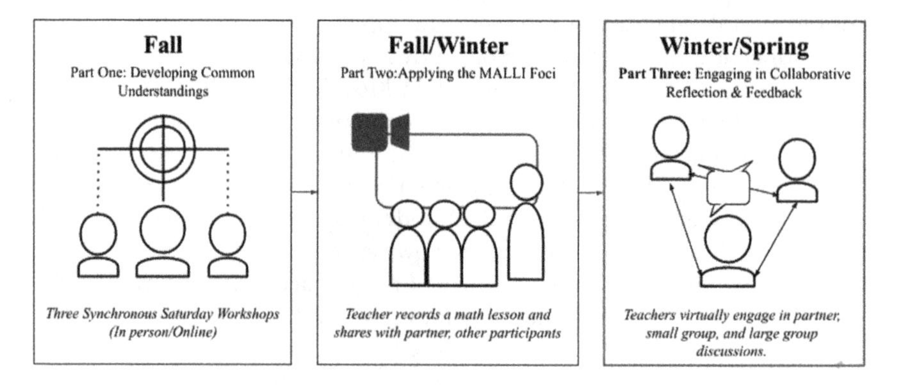

Figure 7.1 The three parts and timeline utilized for LSVC.

lecture-like format. Facilitators provide materials (handouts, slides, activities) and share current research regarding mathematics content learning and meta-language. As a final activity for the first workshop, teachers are given significant time and flexible structures to collaboratively plan potential lessons to record and use for the lesson study component of the TPD.

Subsequent workshops in Part I extend teachers' understandings and applications of the pedagogies. Teachers share classroom artifacts and student work. All teachers proctor a grade-based, common mathematics writing assessment. One workshop provides time and structures for small groups of teachers to calibrate and discuss scoring processes. Another workshop invites mentor teachers and their paired university-based student teachers to discuss MALLI pedagogies and practices with other mentor/mentee partners. These university-based student teachers are concurrently enrolled in coursework that integrates the same MALLI pedagogies. Also, during Part I, LSVC facilitators provide structures for mentor teachers to engage in dialectic feedback conversations based on their observations of their student teacher's MALLI-based lessons. At the conclusion of Part I, teachers create a lesson plan to feature one of the three MALLI pedagogies.

Part II: Applying the MALLI Foci

Once they have a mathematics lesson plan, teachers video record themselves teaching it. They share the lesson with another LSVC participant, who teaches a similar grade. Each partner watches the other's lesson, on their own time. Facilitators provide protocols to support these peer observations. Once both teachers have had the opportunity to record and share their videos with their partner, the teachers conduct their own feedback conversation via video call. The LSVC facilitators support this conversation with a short list of guiding questions. Teachers are also asked to complete a brief written reflection after their partner conversation.

Part III: Engaging in Collaborative Reflection and Feedback

The final part of LSVC brings the pairs of partners from Part 2 together for small group conversations. Each teacher selects and shares a small clip with the group, providing context and often asking a question of the group. Together, they reflect on the watched lessons during the second virtual workshop. A final meeting culminates the project, celebrating the year's learning, with opportunities to share new perspectives, suggestions, and takeaways. Clips are extracted from the videos and the teachers' contributions are compiled into a video library that participants can access once the LSVC TPD concludes.

RESULTS: STATISTICALLY SIGNIFICANT AND QUALITATIVELY SOUND

Quantitative analysis of teachers' pre- and post-survey responses yielded statistically significant evidence that the teachers perceived the combined features of the LSVC hybrid TPD format as effective for their learning and development. Notably: Teachers reported an increase in their perceived confidence to model bilingual instructional practices to preservice and intern teachers. They also reported enacting bilingual instructional behaviors in mathematics classes. Further, the results of qualitative analysis provided evidence for the basis of this perception. Teachers described LSVC as consistent and sustained; easily accessible; a means to access new perspectives of their teaching and of others; collaborative; relevant—the participating teachers sought a means to increase their efficacy in teaching mathematics in DLP environments, a topic not covered by their district/school TPD in-service days; and *different.* The last point suggests that these teachers' prior experiences with TPD was not as effective as their time spent in the LSVC activities.

RECOMMENDATIONS FOR EFFECTIVE, TEACHER-CENTERED TPD IN SPECIALIZED CONTEXTS

Creators and users of TPD should continue to seek models that fulfill an expanded understanding of "school-wide" and "collaboration." Models like LSVC are convenient, consistent, and sustained. Designs that modestly modify research-proven structures, like lesson study, and incorporate components from other effective structures, like video club, have the potential to make TPD more accessible (e.g., Alles et al., 2019), more convenient, and ultimately, more effective for participating teachers. Technology advances, combined with the necessity for teachers to become familiar with video interactions and asynchronous participation, provide an opportunistic environment to transcend geographic boundaries, time limitations, and financial barriers.

Characteristics of Effective Teacher Professional Development

Current research indicates the following features that result in effective TPD: (a) content-focused; (b) collaborative in nature; (c) involves teachers as active learners, provides plenty of opportunities to practice; (d) sufficient modeling of effective practices; (e) sufficient synchronous and asynchronous coaching and support; (f) accessible resources; and (g) enough time and space for teachers to talk with each other (e.g., Darling-Hammond & McLaughlin, 1995; Fishman et al., 2003; Wilson & Berne, 1999). Findings from the LSVC project corroborates this existing literature on the benefits of teachers viewing themselves and others teaching in action through video recordings. The LSVC project expands the field's understanding of how technology can amplify and extend opportunities for experientially and geographically diverse teachers to interact and collaborate. This version of LSVC provided for differentiated and effective learning experiences for teachers at different stages of their career. Additionally, teachers who collaborate on planning lessons and engage in reflective conversations gives those teachers insight into their teaching and student learning (Fernandez & Yoshida, 2004; Takahashi & Yoshida, 2004; van Es & Sherin, 2008; van Es, 2012).

Through the LSVC structure, participating teachers learned through strategically combined elements of the video club and lesson study frameworks. Teachers watched themselves and others teaching in action. They discussed specific actions, classroom dialogue, and visual products. Honing in on specific, observable events is a means to both self-reflection and considerate and critical feedback to peers, a key step in creating a teacher professional learning community (van Es, 2012). TPD structures that begin with a general theme or concept create a flexible conversational space, where teachers from diverse environments with diverse backgrounds can reflect and develop new ideas about their teaching practices and their students. The three-part LSVC structure encourages participants to develop a sustainable network of support, providing enough time to put ideas into action, and timely, consistent gatherings for teachers to reflect and discuss. Further, the non-evaluative nature of the research team and other teachers made it possible for teachers to talk about their concerns and exchange ideas freely.

For the MALLI participants, the hybrid TPD model provides a way to meet other DLP teachers teaching in various DLP programs across states and districts. They watch themselves and others' classroom videos, forming a professional community where they can ask questions, exchange ideas and seek help when needed. The hybrid in-person and online structure creates flexibility for teachers who are physically far apart or whose prep schedules never overlap. The specialized learning needs of DLP teachers are difficult to meet when the pool of faculty is limited to a single school or district—these specialized teachers benefit especially from conversations with other DLP colleagues across the country. The hybrid format also enables teachers to notice student learning, in addition to their teaching practices, a finding consistent with existing literature (Tunney & van Es, 2016; van Es & Sherin, 2008).

The LSVC model is an effective TPD structure because it intentionally guides teachers through dialogue. The active model avoids passive instructional pedagogies and top–down authoritarian approaches. LSVC provides support and structures for teachers to develop their understanding of the specialized content, at their own pace and explicitly connected to their own experiences and current classroom practices. This is an essential characteristic of effective TPD. Researchers, administrators, or both initially, and briefly, provide basic structures and general information. Then, these authority figures *step away* for teachers to discuss and learn from each other. Researchers or administrators wishing to implement effective TPD should fulfill the role of ongoing, consistent, facilitators, rather than experts who "teach" and leave. The model can be especially helpful for teachers who have anxiety about or lack confidence/resources to teach mathematics, a common concern among female (e.g., Stoehr, 2017), elementary (Brown et al., 2012), and multilingual teachers (Palmer et al., 2016). The collaborative nature of LSVC, combined with a specific focus (choosing literacy, language, or discourse) in the explored model, supports teachers with these traits.

LSVC teachers report that the collection of pedagogy foci-based classroom videos are valuable resources for them in reflecting on their own teaching, allowing them to "steal" ideas and strategies for their own classrooms. In addition to the provision of practical resources, sustainability is another essential characteristic of effective TPD (Darling-Hammond et al., 2017). Effective PDs need time to be translated to effective practice in classrooms. Sustaining meaningful interactions requires the flexible use of a variety of mediums to deliver and to participate in PD, including face-to-face, video and or/audio, online/print, and various combinations. The medium used to deliver TPD is a significant factor in how effective the TPD is for participants (Fishman et al., 2003). The combination of lesson study and video club used in MALLI created a "user-friendly" format of PD for teachers across different states to make full use of the resources at their convenience.

Sustainability and the mediums used to enact TPD are meaningless if structures are not inherently collaborative. Teachers' learning and ability to change their practice happens in and through the interactions and exchange of ideas among teachers (Little, 2002; Wilson & Berne, 1999). Teachers in MALLI were given consistent and semi-structured opportunities to interact with teachers beyond their school walls. The model was especially helpful for teachers in multilingual mathematics classrooms and early elementary teachers who believed their personal pedagogical expertise was language and literacy. These teachers often discussed how it was challenging to identify connections between mathematics learning and developing language skills. These same teachers often identified elevated language and literacy practices in their *partner's* teaching videos, before seeing or identifying them in their own lessons. Being able to see mathematics taught, in different environments and by different educators, enhanced the participants' ability to learn and elevate pedagogical practices in their own classrooms.

The LSVC TPD structure provided opportunities for consistent "turn and talk" partners, as well as smaller group conversations, and whole-group shares, across grade levels and years' experience. LSVC teachers enjoy having partners from the same state or another state, because they learn about other classrooms, state standards and instruction, as well as relate to personal stories told through in-person and virtual conversations. In addition, MALLI teachers reported that "seeing" the lessons via videorecording was crucial to understanding the mathematics instruction. Generally, teachers learning mathematics pedagogies need video recordings or other direct observation to visualize how manipulatives or other teaching tools are used. Previous research has clearly shown that teachers "notice" details in mathematics instruction when watching video recordings (e.g., Van Es & Sherin, 2008). In fact, it is possible that teachers learning new methods in the language arts could learn from a written transcript of a lesson. This does not appear to be the case for mathematics instruction, although more research is needed to verify this hypothesis.

Another key feature of this effective model is that it sets a foundation for developing long-term professional learning communities. Teachers gain access to collaborative collegiate relationships that transcend practical and geographical boundaries. While walking over to a neighboring classroom to talk over a problem of practice is usually an option, the LSVC model makes it possible to have similar sustained relationships with teachers across districts and the country. This hybrid TPD model can be applied to various programs and classrooms, for teachers teaching different subjects in different languages, grades, and school structures. For under-resourced schools and districts, the LSVC model provides a feasible and affordable structure for teachers to establish and maintain a professional learning community.

For school administrators or district curriculum leaders considering using LSVC or other hybrid TPD models, it might be difficult to part with the traditional in-service activity of "bringing in an expert." But this time-honored strategy is largely what gave TPD such a negative reputation in the first place. When teachers meet their learning needs in a TPD that is enjoyable and useful (e.g., Brenneman et al., 2018), they are more likely to effectively and consistently implement TPD content. Comments from participants in the LSVC model overwhelmingly suggest that teachers seem to prefer to study and improve their pedagogy *without* the assistance of experts or administrative leadership.

Centering Participants' Specialized Needs, Leaning Into Differentiation

Dual language programs continue to expand across the United States, nearly tripling from 2011–2021 (American Councils Reacher Center, 2022). This rapid expansion in format necessitates parallel expansion in TPD. DLPs are spread across the nation and are not often located in the same district, much less the same county or state. There is a significant need for DLP teachers to connect and share experiences, to learn from each other's successes as well as challenges and obstacles (Alvaryero Ricklefs, 2022). The specialized nature of these programs is rife with creative teaching and learning opportunities that too often occur in centralized silos. Models like LSVC, that expand beyond the reaches of one school or one district and develop decentralized learning communities (Moloney & Wang, 2016), enable cross-pollination of ideas and provide opportunities for teachers to develop meaningful professional relationships. Collaborative and interactive models are critical for DLP teachers working to develop efficacy in teaching mathematics while holding constant attention to language. Seeing effective mathematics teaching in other classrooms and speaking directly with other DLP mathematics teachers about problems of practice diminished teachers' anxiety and increased feelings of pedagogical efficacy in the LSVC study discussed above. Simply reading an article or watching a webinar does not have the same level of personalized connection or ability to build professional relationships that the LSVC model offers.

A focus on specialized content and applicability to the unique teaching experiences and contexts of the participating teachers are two of seven critical features of effective TPD (Darling-Hammond et al., 2017). LSVC teachers are motivated to participate in the TPD because of the specialized content, the structure and delivery of the content, and relevant opportunities to practice what they learn. When TPD content is relevant to what teachers teach in their classrooms, is in alignment with their curricula and school requirements, and provides the opportunity for teachers to study an isolated but multifaceted element of pedagogy in a content area, there is usually a connection to the perceived effectiveness of teacher PDs (Darling-Hammond et al., 2017).

Hybrid models, like LSVC, allow DLP teachers to interact beyond the walls of their home school, and provide the means for these specialized teachers to develop substantive professional networks relevant to their needs. For example, in DLP programs the focus is understandably on language; however, the connection between language and mathematics often gets lost. Teachers in these environments look for ways to eliminate the either/or binary of mathematics/language content areas and seek to approach the content areas from a both/and perspective. Bringing this connection to the forefront can be achieved by expanding the conversation outward from a single school. The practical concerns of when to meet and where are critical features of TPD programs that seek to support teachers in specialized contexts, like DLPs (Brenneman et al., 2018). Expanded networks are not feasible without extending the definition of "school-wide," and abandoning the requirement for face to face, synchronous conversations.

TPD that is able to effectively differentiate across a diverse body of teachers, for example TPD that can simultaneously address teachers' similarities (e.g., teaching in a DLP) and differences (e.g., various years of experience), is widely considered a crucial aspect of teacher education. The Organisation for Economic Cooperation and Development ([OECD]; 2019) names "supporting the professional growth of teachers and school leaders throughout their careers" as a major goal for educational systems worldwide (p. 38). Further, the OECD suggests this should be accomplished through *easily accessible* TPD *tailored to teachers' levels of experience and specific teaching contexts.* The National Council of Teachers of Mathematics (NCTM) recommends that teachers be mentored *throughout* their teaching careers (NCTM, 2013) and that a primary goal of teacher evaluation is to support professional growth (NCTM, 2016), across the career spectrum. Yet, differentiated TPD opportunities that recognize and attend to teachers' years of experience are limited (Kise, 2017; OECD, 2019).

The different challenges and needs that teachers have, including years' experience, calls for differentiated support in order for teacher learning to happen. Several studies indicate that the different challenges and needs of teachers across the career spectrum calls for differentiated professional development support in order for teacher learning to happen (e.g., Borko, 2004; Brody & Hadar, 2015; Lienert et al., 2001). In specialized contexts like DLPs, teaching and learning challenges are heightened for teachers: not only do colleagues down the hall and in the same school have varied years' experience in dual language classrooms, the required pedagogical content knowledge but the required pedagogical content knowledge is also unique in DLPs. Teachers in the specialized DLP context must also help their students meet the unique, dual goals of biliteracy and bilingualism (Howard et al., 2007).

Models like LSVC could work for various specialized groups of teachers, and facilitators can bring them together, regardless of geographic location. As with most professional disciplines, the work of teachers is now more specialized than ever. LSVC can be a productive vehicle for teachers to acquire such specializations, particularly if it situates teachers as active and agentive learners. One potential limitation of the model is that it relies upon active engagement and motivated participants. Shifting district curricular focus and changing administration can complexify a teacher's willingness to participate in such TPD—it could be argued that the success of LSVC was due to participants who volunteered. However, two features of LSVC mitigate this argument. LSVC, importantly, is an affordable and time-flexible model of TPD for teachers. This is particularly crucial for the many schools facing shrinking budgets and teacher shortages and for the teachers who are finding they have less time to prepare lessons during the school day. Hybrid models, like LSVC, can serve as bridges to connect teachers working in unique contexts who often express feelings of isolation, leading to increased efficacy in teaching, yielding the potential for higher mathematics learning (and assessment scores) for students. Models that emphasize interaction and conversation, through semi-structured synchronous and asynchronous opportunities, allows teachers to share stories about their students and may offer life-long professional communities.

IMPLICATIONS AND CONCLUSION

The continued exploration and implementation of hybrid models is essential to the future of teacher development, especially in their development of mathematics teaching. TPDs like LSVC offer feasible methods for any network of specialized teachers, like those in DLP programs teaching mathematics, to form collaborative relationships that transcend geographical boundaries, enabling them to share recorded lessons, exchange ideas and participate in a sustained, rigorous and rewarding professional learning community. Teachers who are anxious about their mathematics instruction could be supported through the collaborative, non-judgmental, learning-centered professional culture that LSVC strives to embody, improving their sense of efficacy. This is an area worthy of exploration, as it is well documented that female teachers, who comprise the majority of elementary school faculty, have high levels of anxiety around teaching mathematics. Experienced teachers know that their students provide the best measure of the quality of their instruction. So-called instructional experts leading TPD will never eclipse what professional teachers gain by carefully analyzing their own students' understandings. Studies of the LSVC hybrid model have confirmed this finding but have also shown the importance of developing communities for teachers to share and discuss actual mathematics instruction with other teachers who share similar challenges.

REFERENCES

Alles, M., Seidel, T., & Gröschner, A. (2019). Establishing a positive learning atmosphere and conversation culture in the context of a video-based teacher learning community. *Professional Development in Education, 45*(2), 250–263. https://doi.org/10.1080/19415257.2018.1430049

Alvayero Ricklefs, M. (2023). Supporting novice dual-language teachers in the U.S. midwest. *The International Journal of Learning in Higher Education. 30*(2), 77–96. https://doi.org/10.18848/2327-7955/CGP/v30i02/77-96

American Councils Research Center. (2022). 2021 canvass of dual language and immersion (DLI) programs in U.S. public schools, American councils for international education. Retrieved in May 2023 from https://www.americancouncils.org/sites/default/files/documents/pages/2021-10/Canvass%20DLI%20-%20October%202021-2_ac.pdf

Bigsby, J. B., & Firestone, W. A. (2017). Why teachers participate in professional development: Lessons from a schoolwide teacher study group. *The New Educator, 13*(1), 72–93.

Borko, H. (2004). Professional development and teacher learning: Mapping the terrain. *Educational Researcher, 33*(8), 3–15.

Borko, H., Jacobs, J., Eiteljorg, E., & Pittman, M. E. (2008). Video as a tool for fostering productive discussions in mathematics professional development. *Teaching and Teacher Education, 24*(2), 417–436.

Brantlinger, A., Sherin, M. G., & Linsenmeier, K. A. (2011). Discussing discussion: A video club in the service of math teachers' national board preparation. *Teachers and Teaching: Theory and Practice, 17*(1), 5–33.

Brenneman, K., Lange, A., & Nayfeld, I. (2019). Integrating STEM into preschool education; designing a professional development model in diverse settings. *Early Childhood Education Journal. 47*, 15–28. https://doi.org/10.1007/s10643-018-0912-z

Brody, D. L., & Hadar, L. L. (2015). Personal professional trajectories of novice and experienced teacher educators in a professional development community. *Teacher Development, 19*(2), 246–266.

Brown, A., Westenskow, A., & Moyer-Packenham, P. (2012). Teaching anxieties revealed: Pre-service elementary teachers' reflections on their mathematics teaching experiences. *Teaching Education, 23*(4), 365–385. https://doi.org/10.1080/10476210.2012.727794

Bunch, G. C. (2013). Pedagogical language knowledge: Preparing mainstream teachers for English learners in the new standards era. *Review of Research in Education, 37*(1), 298–341. https://doi.org/10.3102/0091732X12461772

Cajkler, W., Wood, P., Norton, J., & Pedder, D. (2014). Lesson study as a vehicle for collaborative teacher learning in a secondary school. *Professional Development in Education, 40*(4), 511–529.

Coenders, F., & Verhoef, N. (2019). Lesson Study: Professional development (PD) for beginning and experienced teachers. *Professional Development in Education, 45*(2), 217–230.

Darling-Hammond, L., Hyler, M. E., & Gardner, M. (2017). *Effective teacher professional development.* Learning Policy Institute.

Darling-Hammond, L., & McLaughlin, M. W. (1995). Policies that support professional development in an era of reform. *Phi Delta Kappan, 76*(8).

Feiman-Nemser, S. (1983). Learning to teach. In L. Shulman & G. Sykes (Eds.), *Handbook of teaching and policy* (pp. 150–170). Longman.

Fernandez, C., & Yoshida, M. (2004). *Lesson study: A Japanese approach to improving mathematics teaching and learning.* Routledge.

Fishman, B. J., Marx, R. W, Best, S., & Tal, R. T. (2003). Linking teacher and student learning to improve professional development in systemic reform. *Teaching and Teacher Education, 19*(6), 643–658. https://doi.org/10.1016/S0742-051X(03)00059-3

Fullan, M., & Hargreaves, A. (2016). *Bringing the profession back in: Call to action.* Learning Forward.

Galguera, T. (2011). Participant structures as professional learning tasks and the development of pedagogical language knowledge among preservice teachers. *Teacher Education Quarterly, 38,* 85–106.

Howard, E. R., Sugarman, J., Christian, D., Lindholm-Leary, K. J., & Rogers, D. (2007). *Guiding principles for dual language education.* Center for Applied Linguistics.

Irby, B. J., Sutton-Jones, K. L., Lara-Alecio, R., & Tong, F. (2015). Informal individual learning via virtual professional development: A proposal for massive open online professional informal individual learning (MOOPIL). In F. Nafukho & B. Irby (Eds.), *Handbook of research on innovative technology integration in higher education* (pp. 343–355). IGI Global.

Kelly, P. (2006). What is teacher learning? A socio-cultural perspective. *Oxford Review of Education, 32*(4), 505–519.

Kise, J. A. (2017). *Differentiated coaching: A framework for helping educators change.* Corwin Press.

Kizilbash, Z. (2020). How teachers experience learning and change: A phenomenographic study of internationalized teacher professional development. *Teacher Learning and Professional Development, 5*(1), 1–14.

Lee, O., Adamson, K., Maerten-Rivera, J. L., Lewis, S., Thornton, C., & Leroy, K. (2008). Teachers' perspectives on a professional development intervention to improve science instruction among English language learners. *Journal of Science Teacher Education, 19*(1), 41–67. https://doi.org/10.1007/s10972-007-9081-4

Lienert, C., Sherrill, C., & Myers, B. (2001). Physical educators' concerns about integrating children with disabilities: A cross-cultural comparison. *Adapted Physical Activity Quarterly, 18*(1), 1–17.

Little, J. W. (2002). Professional community and the problem of high school reform. *International Journal of Educational Research, 37*(8), 693–714.

Llosa, L., Lee, O., Jiang, F., Haas, A., O'Connor, C., Van Booven, C. D., & Kieffer, M. J. (2016). Impact of a large-scale science intervention focused on English language learners. *American Educational Research Journal, 53*(2), 395–424. https://doi.org/10.3102/0002831216637348

Lynch, J., Irby, B. J., Tong, F., Lara-Alecio, R., Zhou, Z., & Singer, E. (2021). Massive open online professional individualized learning: building teachers' instructional capacity for English learners. *Teaching English as a Second or Foreign Language, 25*(2).

Maldonado, S. I., Mosqueda, E., Bravo, M. A., & Solís, J. L. (2021). Assessing and teaching students' biliteracy in mathematics: A professional development model. *The Multilingual Educator,* 36–39.

Marks, H. M., & Louis, K. S. (1999). Teacher empowerment and the capacity for organizational learning. *Educational Administration Quarterly, 35*(5), 707–750.

Moloney, R., & Wang, D. (2016). Limiting professional trajectories: A dual narrative study in Chinese language education. *Asian Pacific Journal of Second and Foreign Language Education, 1*(1). https://doi.org/10.1186/s40862-016-0007-6

National Council of Teachers of Mathematics. (2013). *Teacher mentorship: A position of the national council of teachers of mathematics.* https://www.nctm.org/uploadedFiles/Standards_and_Positions/Position_Statements/Teacher%20Mentorship.pdf

National Council of Teachers of Mathematics. (2016). *Evaluation of teachers of mathematics: A position of the national council of teachers of mathematics.* https://www.nctm.org/uploadedFiles/Standards_and_Positions/Position_Statements/Teacher%20Mentorship.pdf

OECD. (2019). The consequences of industry 4.0 for the labour market and education. *The Future of Education and Labor,* 37–56.

Palmer, D., Henderson, K., Wall, D., Zúñiga, C. E., & Berthelsen, S. (2016). Team teaching among mixed messages: Implementing two-way dual language bilingual education at third grade in Texas. *Language Policy, 15*(4), 393–413. https://doi.org/10.1007/s10993-015-9361-3

Putnam, R. T., & Borko, H. (2000). What do new views of knowledge and thinking have to say about research on teacher learning. *Educational Researcher, 29*(1), 4–15.

Shulman, L. S. (1987). Knowledge and teaching: Foundations of the new reform. *Harvard Educational Review, 57*(1), 1–22.

Stoehr, K. J. (2017). Mathematics anxiety: One size does not fit all. *Journal of Teacher Education, 68*(1), 69–84. https://doi.org/10.1177/0022487116676316

Takahashi, A., & Yoshida, M. (2004). Lesson-study communities. *Teaching Children Mathematics, 10*(9), 436–437.

Thompson, A. (2008). *Using video technology to provide a professional development forum for reflection on the use of academic language for mathematics in elementary school teachers* [Conference presentation]. The California Mathematics Council North, Asilomar, CA.

Tong, F., Luo, W., Irby, B. J., Lara-Alecio, R., & Rivera, H. (2017). Investigating the impact of professional development on teachers' instructional time and English learners' language development: A multilevel cross-classified approach. *International Journal of Bilingual Education and Bilingualism, 20*(3), 292–313. https://doi-org.oca.ucsc.edu/10.1080/13670050.2015.1051509

Tunney, J. W., & van Es, E. A. (2016). Using video for teacher-educator professional development. *The New Educator, 12*(1), 105–127. https://doi.org/10.1080/1547688X.2015.1113348

van Es, E. A. (2012). Examining the development of a teacher learning community: The case of a video club. *Teaching and Teacher Education, 28*(2), 182–192. https://doi.org/10.1016/j.tate.2011.09.005

van Es, E. A., & Sherin, M. G. (2008). Mathematics teachers' "learning to notice" in the context of a video club. *Teaching and Teacher Education, 24*(2), 244–276. https://doi.org/10.1016/j.tate.2006.11.005

Wilson, S. M., & Berne, J. (1999). Teacher learning and the acquisition of professional knowledge: An examination of research on contemporary professional development. *Review of Research in Education, 24*, 173–209. https://doi.org/10.3102/0091732X024001173

Xia, Y., Patthoff, A., Bravo, M., & Téllez, K. (2022). "We don't observe other teachers": Addressing professional development barriers through lesson study and video clubs. *Teacher Learning and Professional Development,* 7(1), 1–22. https://journals.sfu.ca/tlpd/index.php/tlpd/article/view/88

CHAPTER 8

"AHORA YA SÉ QUÉ HACER"

How Translanguaging Mediates Bilingual Teacher Candidate Reflections and Teaching of Mathematics

Jorge L. Solís
University of Texas San Antonio

Brenda Sarmiento-Quezada
Purdue University

Lina Martin Corredor
Metropolitan State University at Denver

ABSTRACT

This chapter examines how translanguaging mediates bilingual teacher candidate (BTC) reflections and their teaching of mathematics lessons in dual language contexts. Mediation is the process by which symbolic and material tools or artifacts are used to make sense of, learn, and act on/in the world (Moll, 2013; Wertsch, 1991). For BTCs, translanguaging as mediation, refers to how BTCs learn to teach mathematics to and through translanguaging. Mediation

Mathematics Instruction in Dual Language Classrooms, pages 135–155

is an essential concept in the preparation of more effective and inclusive bilingual teachers (Martínez-Álvarez, 2020). In this case study of BTC learning to teach mathematics in bilingual contexts, we examine how translanguaging is used as a tool for learning and becoming a bilingual teacher and for teaching K–5 students in the context of mathematics. Moreover, teacher reflection is a critical stance-taking practice connected to culturally and linguistically responsive teacher identities (Galindo, 1996; Martin & Strom, 2016). This analysis draws from data collected from a cohort of BTCs involved in the Mathematics and Language, Literacy Integration (MALLI) project which was a longitudinal 5-year project that took place in Texas and California.

There are several principles driving this focus. First, we consider translanguaging a ubiquitous naturally occurring communicative phenomena fundamentally part of the language of bilingual interlocutors (teachers, students, families) and within social contexts (schools, home, public spaces) including STEM contexts (Langman et al., 2021; Solís et al., 2018). Second, translanguaging, as a distinctive aspect of bilingual discourses, is shaped by and shapes larger socio-historical-political forces and ideologies. Third, translanguaging is part and parcel of the production and reproduction of sociocultural-sociolinguistic knowledge. Fourth, translanguaging for teachers implicates two overlapping aspects of communicative social action including shifts between languages (*doing translanguaging*) and referencing shifts in language (*talking about translanguaging*).

THE MALLI PROJECT AND STUDY

The MALLI project aimed to prepare dual language bilingual teacher candidates (BTC) to integrate both lesson study (LS) professional development stances and language-rich mathematical practices for the benefit of K–5 dual language learners, teachers, and families. The MALLI framework incorporates more expansive mathematical disciplinary teaching and learning approaches to mathematical biliteracy development. MALLI teaching practices are intended to help researchers, BTCs, teacher mentors, and teacher educators assess, develop, support, and promote: (a) mathematics discourse, (b) mathematics bi/literacy, and (c) mathematics vocabulary. MALLI practices are interconnected in how bilingualism and biliteracy can be promoted in mathematics that hinges on more expansive notions of mathematics and language. Mathematical biliteracy considers how texts, words, and discourses are leveraged dynamically and transcultured within and across languages and spaces in the service of mathematical sense-making (Celedón-Pattichis et al., 2012; García et al., 2014; Guerra, 2015; Pierson et al., 2021; Uribe-Flórez et al., 2014; Vomvoridi-Ivanovic, 2012; Yeh, 2017; Zavala 2017). Table 8.1 summarizes the way MALLI conceptualized

TABLE 8.1 Overview of MALLI Teaching Practices

MALLI Practice	Definition	Examples of Strategies & Activities
Mathematics Discourse	Talking and acting to accomplish mathematics practices such as proving or explaining mathematics solutions, problems, or statements	• Linking home/community funds of knowledge • Design of interactive classroom spaces for collaboration • Cultivate mathematics arguments and explanations • Focused mathematics discourse and talk
Mathematics Biliteracy	Attention to reading and writing in mathematics including discussions and interpretations of math texts and/or how to produce/read different types of mathematics texts	• Expanded mathematics texts including diaries, learning logs • Data based reflections, observations • Peer review • Anchor/mentor word problems
Mathematics Vocabulary	Attention to the special meanings of words used in mathematics and how to learn and reinforce specialized and precise meanings through the use of background knowledge, morphology, cognates, collocations, and noun phrases	• Analyze and classify keywords in a mathematics texts including familiar & familiar words; tier 1, 2, 3 words • Attend to disciplinary idioms, roots, suffix, prefix, cognates, collocations, noun phrases, nominalization

the central pedagogical practices incorporated into MALLI's BTC professional development contexts (university courses, LS meetings, coaching).

The MALLI pedagogical domains were incorporated into the curriculum of two required PK–5th grade bilingual teacher preparation courses including bicultural-bilingual approaches to teaching mathematics and bicultural-bilingual approaches to integrating content area instruction. Six bilingual anchor MALLI framework lessons were developed by the research team and used in these two courses by university instructors. These embedded lessons were taught primarily in Spanish and took place during the semester preceding BTC's clinical teaching placement. During their clinical teaching, BTCs were placed in bilingual classrooms, which included mostly two-way dual language classrooms. Additionally, each BTC was paired with a bilingual mentor teacher who engaged MALLI practices through project-related collaborative coaching sessions.

RELATED RESEARCH AND THEORETICAL PERSPECTIVES

A central premise driving our analysis is the idea that bilingual learners benefit from drawing more fully from their sociocultural repertoires through critical collaborative dialogue (e.g., Gort & Sembiante, 2015). As was the case with our participants, BTCs were prompted to use (and used) their bicultural-bilingual resources when they participated in MALLI activities. To cultivate a collaborative professional learning community among BTCs and between BTCs and their mentors, the MALLI framework incorporated and adapted LS principles (Lewis & Perry, 2014) to engage bilingual teachers in the co-construction of mathematics lessons (see Figure 8.1).

STEM Bilingual Teacher Preparation

Limited research examines the connection between language and mathematics in linguistically diverse classrooms, particularly in bilingual teaching. Novice bilingual teachers receive insufficient attention, facing challenges due to methodological and theoretical dilemmas. Mathematics teaching opportunities for them are often limited and occur irregularly or at the end of their field experience, while reading and writing receive

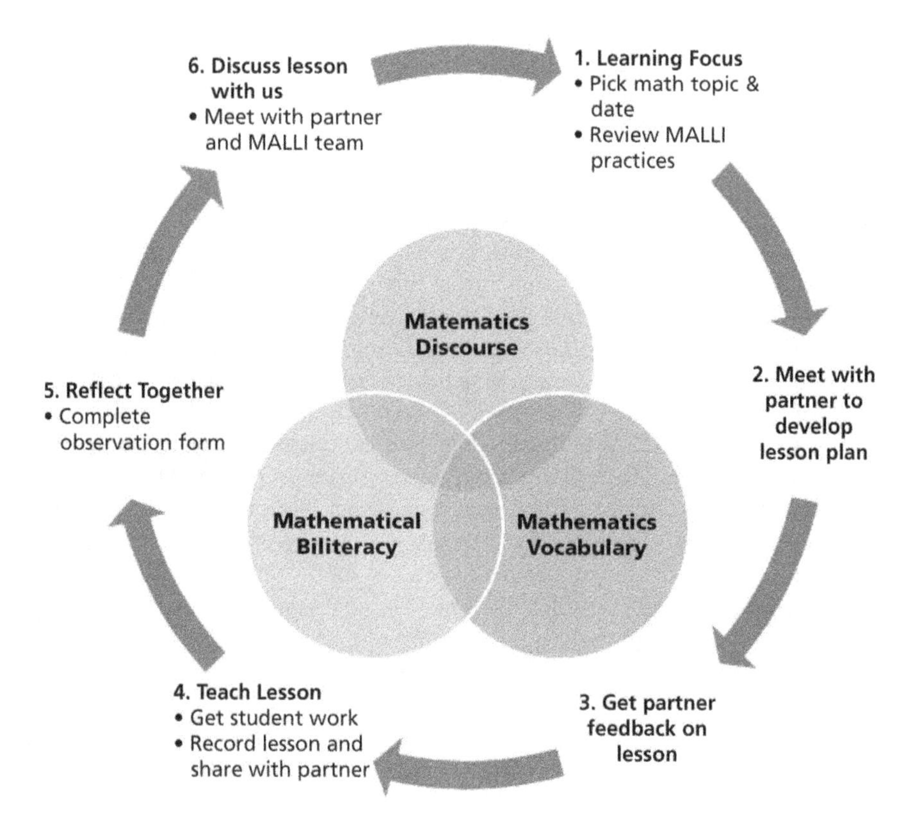

Figure 8.1 BTC lesson study approach.

more focus. Even in dual language models, mathematics and science are primarily taught monolingually in English, reducing chances for Spanish or bilingual instruction (Bravo et al., 2014). The mathematics education community has shown increased interest in bilingual learners, emphasizing the importance of language in STEM education (Moschkovich, 2006; Solís & Bunch, 2016). Research highlights the importance of supporting novice teachers as learners in diverse mathematics classrooms, reflecting on their prior knowledge and experiences (Zavala, 2017). Innovative teacher education programs can foster equity-oriented teacher identities in mathematics by creating opportunities for critical reflection and recognizing the linguistic and cultural knowledge of novice bilingual teachers.

Translanguaging

The interest in bilingual teaching preparation has led to a focus on translanguaging as a pedagogy, recognizing the multiple language practices of bilinguals (Garcia, 2013; Gort & Sembiante, 2015; Musanti & Rodríguez, 2017). Translanguaging can be spontaneous, occurring naturally in everyday contexts, or pedagogical, involving instructional strategies that incorporate multiple languages. Various studies explore how multilingual learning contexts expand linguistic repertoires and support emergent bilingual children, emphasizing the co-construction of discursive spaces in the classroom (Cenoz & Gorter, 2020; Garcia et al., 2012). Translanguaging pedagogy can have transformative effects by improving student language awareness and attitudes toward language learning (Martínez-Álvarez, 2017; Tian & Zhang-Wu, 2022). Moreover, pedagogical translanguaging practices can support synergies between multimodal-multilingual resources and disciplinary STEM practices leading to more inclusive learning environments for bilingual/multilingual students (Pierson & Grapin, 2021).

Lesson Study

Figure 8.1 describes how we adapted the LS model to enhance professional learning among BTCs and their use of mathematical biliteracy practices. The LS model was employed in the MALLI project to promote teacher candidates' learning about mathematics and language integration in the instruction of emergent bilinguals. LS is a proven professional development approach in mathematics education across K–12 settings (Lewis & Perry, 2014; Lewis et al., 2009). The LS model involves a collaborative and iterative process wherein teachers collectively analyze learning challenges, research, create, and revise lesson activities, implement them while observed by peers, and evaluate student work samples and related data. This inquiry cycle emphasizes understanding *how students learn* disciplinary knowledge rather than solely focusing on *what they learn.*

Less is known about how novice bilingual teachers engage in and benefit from their involvement in LS in teacher education programs (see Figure 8.1; Bjuland & Mosvold, 2015; Druken et al., 2021). Novice bilingual MALLI teachers in this study drew from teacher education course strategies and activities aimed at integrating MALLI practices in their mathematics teaching to participate in two LS cycles during their clinical practicum. Initial observations of LS activities indicated that novice MALLI teachers frequently translanguaged during LS and pulled from their cultural and linguistic repertories more fully. From this perspective and inspired by a sociocultural theoretical lens, our research team engaged in a case study analysis focused on the following research question: "How does translanguaging mediate BTC LS reflections and their teaching of mathematics lessons in dual language contexts?"

METHODOLOGY AND DATA ANALYSIS

This case study focuses on 13 BTCs participating in the MALLI project. Their reflective conversations during the LS cycles were transcribed and coded using qualitative data analysis software. The research question for this study aims to understand how novice bilingual teachers reflect, through traslanguaging, on their teaching experiences and emerging bilingual teacher identities. The researchers grappled with clarifying their roles at different points in the project; their roles shifted from being a university course instructor, lesson study collaborators, and a liaison/collaborator with local dual language school mentor teachers and university supervisors. As bilingual educators, we attempted to design activities and materials that were inclusive of the range of BTC voices and experiences and considered this adaptive approach an important part of our bilingual research methodology. We made MALLI BTCs aware of our bilingualism and explicitly welcomed the use of both languages in our discussions throughout our informal communications, development of materials, and formal coaching exchanges. As researchers, we also made participants aware that their participation and commentary was protected from being used to evaluate their standing in the teacher credential program and school district identity despite working closely with their mentor teachers and university supervisors.

BTCs, all Latinx/Hispanic females, engaged in at least two reflections coming at the end of each LS cycle (see Step 5 and 6 in Figure 8.1) and each engaged in two LS cycles (four reflections per participant) that highlighted the benefits of observing their own video-recorded lessons and receiving peer feedback. Researchers invited BTCs to engage in reflective conversations in English, Spanish, or both languages during post-lesson reflections. The questions focused on helping BTCs reflect on the lesson development process and implementation including noticing evidence of student learning and related instructional decisions and activities. The following questions are two examples of questions discussed in the post-lesson reflection.

- Explain which mathematics skill/content area concept and MALLI practice you decided to focus on? How did you decide?
- Did you have any challenges regarding language development and/or content comprehension?

Upon completing our initial coding analysis, we observed frequent occurrences of translanguaging-related themes in the data corpus. These themes encompassed aspects such as managing two languages in teaching, utilizing bilingual materials/tools, and promoting biliteracy in mathematics. To answer our research question, our analysis focused on instances where BTCs used and/or referenced translanguaging while reporting their own speech or the speech of others (e.g., children, mentors, partners, administrators, parents, instructors). Through this examination, we identified specific instances of where BTCs translanguaged and/or talked about translanguaging as part of their reflective narratives.

FINDINGS

Analysis of BTC LS Reflections

Our analysis of LS debriefs with LS pairs shows that BTCs often used translanguaging to describe their instructional focus as they reflected on their MALLI lessons. We facilitated debriefs with BTCs over the course of one academic year and collected reflective conversations between pairs/groups that were transcribed and analyzed. We coded all LS debriefs for the 13 participants using a grounded and project construct-driven analytical approach. Our seven major themes included MALLI practices, learning focus, bilingual materials, discourse, peer/self-feedback, power and voice, and teaching approach (see Table 8.2). Bolded sub-themes were the most common sub-themes within each major theme. What do BTCs reflections tell us about teaching mathematics bilingually? There is no clear progression in how BTCs reflected on teaching mathematics bilingually; however, for some BTCs translanguaging, as a pedagogical orientation and/or reflective communicative resource, was related to their engagement of MALLI practices. A decrease in translanguaging corresponded with a similar decrease

TABLE 8.2 Themes Associated With Translanguaging

Themes	Explanation	Sub-Themes
Discourse (25%)	Discursive resources and devices used for reflection and to tell a story	• Gesture • Questioning/Critical Stance • Reported Speech • Repetition • (translanguaging)
Power and Voice (21%)	What and how identity stances were constructed, shared, accepted, or rejected	• Agentive/Advocacy Stance • Framing Self as Individual/Collective • Self-Affirming • Self-Critical • Praising Partner • Epistemic Stance (knowledge)
Mathematics Learning Focus (17%)	Reference to a particular teaching and/or learning focus	• Developing Two Languages • Higher Order Thinking/Anti-Deficit • Math Skills (concepts/procedures) • Math As Language-Free • Student Engagement • Learning Challenge • Test Preparation • Link to State Standards
Peer/Self Feedback (14%)	Respond to or giving feedback to self or someone else/partner	• Evaluate Changes • Follow-up • Language of Feedback • Positive Link to Self/Own Teaching • Focus on Changing Pedagogy
Teaching Approach (9%)	Attention to or evaluation of teaching approach	• Use of Prior Knowledge/FOK • Working on Future Changes • Managing L1/L2 in class • Self-Rating/Evaluation (1-5) • Explicit Teaching Strategies
MALLI Practices (9%)	Reference to MALLI Practices or related project resources	• Mathematics Biliteracy/Literacy • Mathematics Discourse • Mathematics Vocabulary
Bilingual Mathematics Resources/ Materials (4%)	Identification, creation, and/ or evaluation of bilingual math resources	• School/District Texts • Mentor Created/Supplied Materials • Online Sites/Sources • Self-Crated/DIY • Spanish language Materials

in attention and engagement in MALLI practices during post-lesson debrief reflections.

Translanguaging became a more explicit and intertwined theme in our coding of the first series of debrief LS conversations (cycle 1, cycle 2) where BTCs engaged in *talk about* mediating and managing bilingual mathematics learning through translanguaging strategies. Moreover, while BTCs were invited to engage in the LS group debrief conversations in either Spanish or English, BTCs participation displayed an unexpected and characteristically fluid bilingual ease that extended from previous bilingual conversations; this *use* of translanguaging reflected previous styles of communication started months and years before as part of informal and formal conversations between BTCs and between BTCs and MALLI researchers stemming from bilingual program coursework activities and other one-one-one mentoring conversations. Translanguaging however was part of our initial coding approach that corresponds also to many ideas and most directly to ideas about managing two languages in the classrooms, mathematics biliteracy, developing two languages, and locating authentic bilingual mathematics materials.

DISCUSSION

During LS reflections, instances when BTCs used or talked about translanguaging, "Discourse (25%)" and "Power and Voice (21%)" were the two themes most associated with translanguaging. The interplay between translanguaging, discourse features, and power/voice suggests that BTCs leveraged a range of sociocultural resources for enacting narratives that replayed significant interactions and memories of being and becoming a bilingual teacher. This interplay also highlights how translanguaging was used to mediate (express, play out/with, rehearse, negotiated share, accept, and reject) emerging identities and circumstances; BTCs also used translanguaging to create spaces of vulnerability that appear to be moments where they are being self-critical about themselves. However, we see that these are also instances where BTCs are voicing questions seeking to understand something new or difficult. This also suggests that translanguaging, as an action that draws more fully from BTCs linguistic repertoires, afforded BTCs opportunities to share thinking, feeling, and ideas more openly about themselves with others. We also acknowledge and embrace the fact that as bilingual educators, collaborators, and researchers we were able to support more expansive bilingual reflections than otherwise possible if we were not bilingual or enforced a separation of languages in our discussions.

The other five thematic categories include "Mathematics Learning Focus (17%)," attention to "Peer/Self-Feedback (14%)," articulation of "Teaching Approach (9%)," reference to "MALLI Practices (9%)," and in relation to locating appropriate "Bilingual Mathematics Resources/Materials (4%)." Within each category there were specific areas of emphasis that offer insight into the sociocultural work associated with translanguaging by BTCs during their post-lesson LS reflections. Under Discourse features, *quoted speech/reported speech* represented most of these examples; this feature is the subject of another analysis that proposes a novel theoretical and methodological approach about the inherent and ubiquitous features activated through bilingual teacher narratives. Translanguaging was used commonly with exchanges that expressed a *critical stance* or questioned a common BTC teaching context or policy. Under the Power and Voice category, BTCs most often expressed a degree of vulnerability and being *self-critical* about their teaching, learning, or engagement. Translanguaging was most often used within the category of peer/self-feedback to describe the *language of feedback* used within LS BTC meetings. Moreover, mention of specific *teaching strategies* and *managing two languages in class* were the most common sub-themes under the major category of teaching approaches. *Mathematics vocabulary* was the most referenced MALLI practice while *Mathematics Biliteracy/Literacy* was the least cited MALLI practice. Lastly, when describing appropriate bilingual mathematics teaching materials while translanguaging, BTCs repeatedly discussed not having sufficient access to these materials and therefore going to *online sites* to download and reference mathematics teaching materials; this was in part related to a scarcity of resources available but also BTC familiarity with online sites that offered more engaging/dynamic materials. Below we explain these findings, organized into two themes: (a) explicit attention to dual language development and biliteracy and (b) critical stance, questioning, and translanguaging.

Theme 1 Explicit Attention to Dual Language Development and Biliteracy

Explicit attention to dual language development and biliteracy in mathematics often involved translanguaged reflections because it involved replaying the language used with DL students in class by BTCs to illustrate a particular pedagogical action. Excerpt 8.1 describes how two BTCs (M and E) discussed how they tried to ensure that children from either a Spanish-language background (SLB) or English-language background ([ELB]; *"los niños de inglés"* or *"estudiantes de español"*) understood their lessons. In this example, M starts by noting that sometimes ELBs don't understand a text or visual requiring that she repeat ideas in English. She says, *"Los niños de ingles no a veces no comprenden tanto"* (English language children sometimes don't understand as much). Both BTCs in this example note that sometimes EBLs benefit from additional support in Spanish or English regardless of the designated curricular DL language. M explains she may offer English language support to EBLs if they have difficulty understanding a lesson. She says this while lifting both arms up with flatted hands in a metaphorical weighing motion where each arm represents a language. This topic prompts E to also share her experience with breaking language separation boundaries to provide support to students as she shifts between Spanish time to English time and back. More specifically E notes that sometimes she'll "put in a little English" for the befit of Spanish language learners/ELBs to helps students jog their memories of key terms and procedures in Spanish. M and E recognize the artificial separation of languages in DL models while teaching and report making decisions to increase student comprehension.

Ln	S	Talk
1	M	**Sí, no entendien todavía como lo visual** *Yes they don't yet understand the visual*
2		**Entonces ya lo digo en inglés para que se den una idea** *Then I say it in English so they have an idea*
3		**en los dos idiomas** *in both languages* ((lifts both arms up in a weighing motion with flattened hands))
4	E	**Igual con compreh-comprensión, eh, hay veces que mis estudiantes** *Same with comprehension sometimes my*
5		**De español como de que-- como ya ven la lección en inglés** *Spanish students like they have already seen the lesson in English*
6		**Y son que como quince minutos en inglés** *And its like fifteen minutes in English*
7		**Tienen un entendimiento mejor de, del inglés** *They have a better understanding of, of English*
8		**Entonces cuando ya les estoy repasando en español como de que se** *So when I'm reviewing in Spanish its like*
9		**sí se confunden** *Yes they get confused*
10		**Um entonces cuando están haciendo su actividad uso las palabras en español** *So when they are doing the activity I use words in Spanish*
11		**y todo les platico en español** *And I talk only in Spanish*
12		**Pero, si luego que todavía ya están como de que batallando un poquito** *But if afterward they are still struggling a little*
13		**Pues ya les empiezo, okay, pues 'acuerden'** *Then I begin okay 'remember'*
14		**Like "when we're doing the measurement now"**
15		**I'll put in a little English, explain in English, um,**
16		**and then review it in Spanish**
17		**Kind of just kind of trying to get them that knowledge from both sides** ((both motion with both hands))
18	M	**Los dos idiomas para que ellos comprendan** *Both languages so they understand*

Excerpt 8.1 Ensuring comprehension with DL students.

This explicit discussion about instructional shifting between languages also leads to E mirroring the shifts in the language of reflection as she relates her story to the LS group. E begins her example with the LS group in Spanish and as she models/demonstrates how she "put[s] in a little English" E retells her story in English. This is both a shift in reporting more accurately her story (recounting past events) but also a shift in her reflective voice (commentary).

In another example in Excerpts 8.2a and 8.2b, A, L, and BS (researcher) discussed a similar issue with building mathematics biliteracy with both groups of students. Here A notes that as opposed to ELB students, SLB

Ln	S	Talk
1	A	**Yeah, for mine, I was ready for them to have the misconception** ((clenches her fist))
2		**on the *ecuación* part** *on the equation part* ((opens her hand and presses it onto the top of the table))
3		**Cause the word problem was worded weird**
4		**so it was like**
5		**"El Señor Lopez tiene dos pedazos de sandía"** *"Mr. Lopez has two pieces of watermelon"*
6		**"y se va a cortar en nueve pedazos"** *"And it it going to get cut into nine pieces"*
7		**Ahora ¿cuántos pedazos de sandía tiene?"** *"Now how many pieces of watermelon does he have?"*
8		**And so, a lot of the kids were like**
9		**well "la ecuación va hacer nueve más dos"** *well "the equation is going to be nine plus two"*
10		**and I'm like, "no, it's not"** ((makes an animated expression as she shakes her head gesturing doubt))
11		**"Porque no va ser nueve más dos?"** *"Why is it not going to be nine plus two?"*
12		**And they were like** ((blinks her eyes twice))
13		**"but it is"** ((makes an exaggerated shocked expression))
14		**And they were like trying to tell me why it was and I let them**
15		**And so it was just that misconception**
16		**And a lot of the Spanish language learners are the ones with those misconceptions that they like**
17		**They see the numbers and they're like, "Oh, let me just put it"**
18		**And sometimes like in other problems, I know I've translated it**
19		**and I was like "think about it now"**
20		**So, I'll say it in English to them**
21		**and they're like, "Oh no I'm wrong"** ((shakes her head, rapidly rolls her eyes and smiles))
22		**Like, "never mind let me go think about it"** ((clenches her fist))

Excerpt 8.2a Replaying word problems.

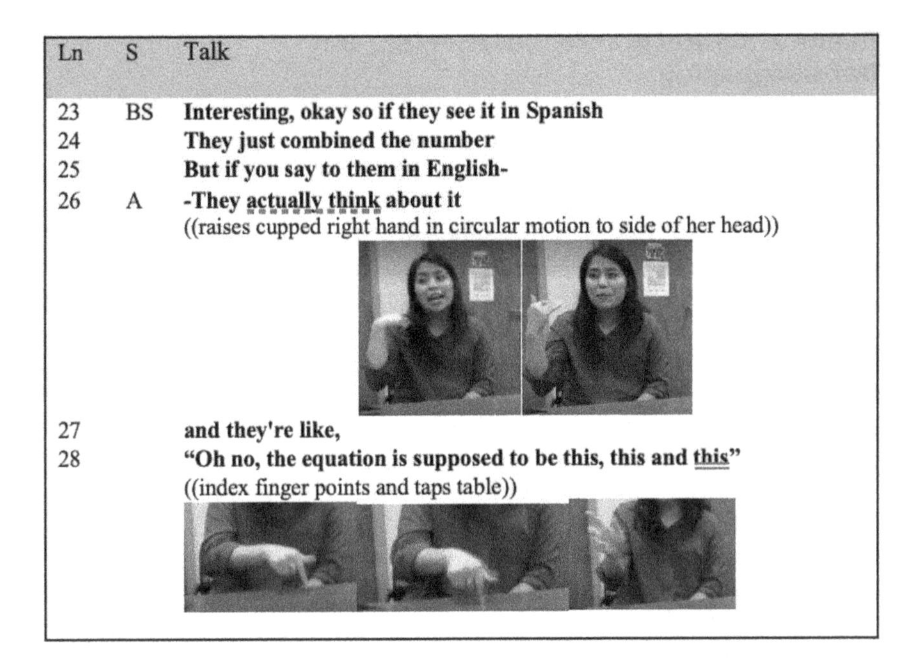

Ln	S	Talk
23	BS	**Interesting, okay so if they see it in Spanish**
24		**They just combined the number**
25		**But if you say to them in English-**
26	A	**-They actually think about it**
		((raises cupped right hand in circular motion to side of her head))
27		**and they're like,**
28		**"Oh no, the equation is supposed to be this, this and this"**
		((index finger points and taps table))

Excerpt 8.2b Gesturing and replaying of word problems.

students don't get easily distracted or confused by word problems in Spanish. She notes that these students often read a word problem and related images and say to themselves, "this is what it's going to be."

Excerpt 8.2a illustrates how BTCs reflected on supporting the mathematics comprehension of DL students recognizing both the need to shift the language of instruction when needed and their interest in assessing the language background of students. Excerpt 8.2b continues the short narrative by BTC A about words problems where she describes through her gestures how she imagines how DL students' mathematics thinking changes (line 26) when she shifts the language of instruction.

BTC A concludes her short narrative by replaying how a particular student responded to solving a word problem displaying the action they took by pointing to the table (line 28) like the student had done after knowing how to write the equation for the word problem.

BTC's reflective narratives detail their initial experiences in bilingual teaching. These narratives prominently feature translanguaging as an indispensable means for situating and molding their accounts. These BTC reflections offer valuable perspectives on how new bilingual teachers formulate and communicate their identities and encounters teaching mathematics bilingually.

Theme 2: Critical Stance, Questioning, and Translanguaging

BTCs also often shift between languages when taking a critical stance or questioning a situation, process, or context. Having a critical advocacy lens for bilingual children and families is a historically significant aspect of being a bilingual teacher. This kind of positionality is a part of the National Dual Language Education Teacher Preparation Standards (Standard 5: Professionalism, Advocacy, and Agency; Guerrero & Lachance, 2018). Our analysis of critical stance draws in part from sociocultural perspectives that grapple with and address social inequities by teachers; for example, a pedagogical "critical stance" is when teachers who "consciously engages learners in interrogating conventional wisdom and practices, reflecting upon ramifications, and seeking actively to transform inequities within their scope of influence in the classroom and larger community" (Teemant et al., 2014, p. 139). Within DL teaching contexts, promoting critical consciousness, and questioning conventional schooling practices by DL educators is important for supporting more equitable treatment of children, families, and communities involved in these programs (Alfaro, 2019; Cervantes-Soon et al., 2017; Flores et al., 2018).

To illustrate how this occurred with BTCs, we describe in excerpt 3 how three BTC participants (M, D, and N) discussed their lesson within the LS group including the university researchers (BS, JS). The discussion at this point of the reflection centers on challenges faced during the lesson especially around mathematics concepts and vocabulary. This MALLI practice (vocabulary) was the most frequently discussed and addressed practice for BTCs; this trend supports previous research with novice teachers as something that is more pedagogically graspable and in line with dominant school-based notions of academic language and literacy development (cite us).

There was general agreement among the three MALLI BTCs that students from different language backgrounds benefitted from differentiation. However, M and D through shifts in language express both critical perspectives and certain personal or professional vulnerabilities in performing up to expectations as a DL teacher. D describes how she didn't encounter many challenges with addressing vocabulary in both languages because she used an anchor chart for her lesson on counting using money units. As an example, she noted that perhaps a "majority English speaker" wasn't picking up all the vocabulary in Spanish; to display understanding this student had said to D, "*Yo tengo un nickel*" which for BTC D indicated that "he was understanding it [the content] but he wasn't learning the vocabulary necessarily." Yet D felt this occurrence was also a plus because this student could be a resource for others as he was understanding the concept for both him and his partner who was also a "majority English speaker." We also see again in this reflection BTC attention to English language background students learning Spanish and less so on Spanish language background students.

Ln	Spkr.	Talk
1	M	**Um, for me it was kind of hard because I could, I mean I was raised**
2		**with like *un penni*, uh, *una moneda*, *un nickle* (*Span. pronunciation*)** *with like a penny, a coin, a nickle*
3		**or whatever**

Excerpt 8.3a Recaling personal history of math terms.

In Excerpt 8.3a (lines 1–3), BTC M picks-up on BTC A's reflection with a contrasting example and experience teaching mathematics vocabulary and related terms in a Spanish language mathematics lesson. BTC M shares repeatedly with the group that the lesson she taught was "kind of hard" for her (line 1) in part because the lesson made her re-examine her family history and that of her students using a variety of terms for units of money in Spanish. M grounds her difficulty in teaching the lesson by noting that she was raised in a bilingual family where family members used a variety of currency terms like "*un penni,*" "*una moneda,*" and "*un knee-kleh*" (Spanish pronunciation). Here M clearly identifies terms that can be and are used in every Spanish language conversation. "Un penni" is an example of the borrowing of the word "penny" or one cent into Spanish. "Una moneda" in Spanish refers to a coin or type of currency but in this case a coin without a specific value. "*Un knee-kleh*" is the borrowing of the word "nickel" or five cents using Spanish language pronunciation. These terms also highlight how some terms are seen as more or less part of standard or academic Spanish language register. This listing of contrasting currency terms by M helps define and historicize her own bilingualism, that of her family, of her students, and how other communities might also use a range of related terms.

Secondly, in Excerpt 8.3b (lines 4–9), M suggests that this ambivalence of using different currency terms was a reason she "would get stuck" teaching the lesson and why she told herself to "get it together." She described

Ln	Spkr.	Talk
4	M	**It was hard for me 'cause I think I you can tell**
5		**And when I was reviewing with them like I would get stuck somewhere**
6		**like, no to get it together "*una peseta*" and I'm like "no"** like, not to get it together "a peseta" and Im like "no"
7		**And I think in one [of] them I do say it**
8		**And then they kept looking at me like what?**
9		**And I was like, "Sí, eso lo decimos pero la manera propia de decirlo es"** *And I was like, "yes we say it like that but the proper way to say it is"*

Excerpt 8.3b Making sense of different math terms.

how in the act of teaching and figuring out which terms to use while attempting to use more standard Spanish terms ("*una peseta*" is an outdated currency term mostly used in Spain), she noticed some students did not recognize some terms like "peseta." M critically notes that she validated students' use of other familiar terms as well but felt the need to tell students that there was a proper way of describing currency terms ("*Si eso lo dicimos pero la manera propia dicir lo es*").

Thirdly, in Excerpt 8.3c (lines 10–17), as M recognized the importance of including student terms while reinforcing more academic terms, she also notes that the lesson was hard for her because she, along with other BTCs, were also corrected by a mentor teacher ("our coach") for using "penni." This recollection from her coach, "*No se llaman en español penni,*" follows a formulaic Spanish-language corrective script seen by bilingual teachers that is both a disapproval of the use of familiar bilingual terms (e.g., penni) and a reinforcement of an idealized, if not pure, Spanish language. This example also illustrates how that Latinx BTCs experience, address, and recall episodes of linguistic violence that questions and attempts to subordinate their bilingual linguistic repertoires and by extension also that of their students (Ek et al., 2013). BTC M is not easily persuaded as she replays the fact that the anchor chart in the lesson uses the "penni" term, and how she adapted the language fitting the pattern in the chart replacing "penni" with "centavo."

Moreover, in Excerpt 8.3d (lines 18–21), M further explained that the lesson was hard for her and likely also for her DL students. She considers that families may also use different terms at home and clarifies to her

4	M	**It was hard for me 'cause I think I you can tell**
5		**And when I was reviewing with them like I would get stuck somewhere**
6		**like, no to get it together "*una peseta*" and I'm like "no"** like, not to get it together "a peseta" and Im like "no"
7		**And I think in one [of] them I do say it**
8		**And then they kept looking at me like what?**
9		**And I was like, "Sí, eso lo decimos pero la manera propia de decirlo es"** *And I was like, "yes we say it like that but the proper way to say it is"*

Excerpt 8.3c Making sense of different math terms.

18	M	**But it was hard for me**
19		**And I can imagine for them if their parents call them that like**
20		**So, I told them**
21		**"Yes, en la casa le puedes decir así pero los que [inaudible] a usarlos en esta manera** *"Yes at home you can say like this but the ones [inaudible] use in this way"*

Excerpt 8.3d Validating home meath languages.

22	M	**It was really hard for me**
23		**For them? I'm assuming yes**
24		**but no, I—the only thing is that the misconception**
25		**the moneda de cinco centavos-- it's bigger so they thought it was more** *the coin of five cents–it's bigger so they thought it was more*
26		**and una moneda de diez centavos is smaller so they thought it was less** *And a coin of ten cents is smaller so they thought it was less*
27		**But overall, I mean they understood it**

Excerpt 8.3e Empathizing with how DL children learn math.

students that it is acceptable to use different terms at home and at school. M connects her previous statements about using and knowing different terms to avoid reproducing inequities in how knowledge from children and families are represented in school classrooms.

BTC M ends her reflective turn in Excerpt 8.3e (lines 22–27) reiterating that teaching this lesson was in fact "very hard for me" and also for students but for another reason. Here again BTC M shifts back to the language of instruction to note that there was a minor misconception between the size of coins used in the lesson and their relative value.

Similar to other BTC reflections, BTC M positions herself as a learner like her students learning mathematics. This dual reference perspective by BTCs is used frequently by BTCs in LS reflections.

CONCLUSION

This case study offers implications for researchers and practitioners engaged in the preparation of bilingual teachers in mathematics. BTC participation in the MALLI project and study afford an examination of how BTCs use their status as teacher candidates and as bilingual educators to engage in inquiry about their own learning of innovative pedagogical practices (MALLI framework) in their teaching of mathematics as well as the mathematics learning of their DL students. This orientation is exemplified in BTCs exchanges like the one by Moira who voiced, during a reflection when thinking through MALLI practices, *"Ahora ya sé qué hacer"* (Now I know what to do). Results from our analysis of lessons study reflections supports the importance of cultivating collaborative bilingual professional learning communities where peers co-construct lessons, offer feedback based on lesson observations, and engage in reflective practices (see Figure 8.1). The MALLI LS cycle is an approach that offers a way for BTCs to be researchers of their learning and teaching. BTC reflections during LS meetings commonly led to BTCs acknowledging each other's expertise, concerns, questions, and experience. This peer and self-validation of becoming a bilingual teacher was expressed repeatedly in LS reflections where BTCs disclosed uncertainties and discomforts while also positioning themselves as critical advocates of teaching mathematics bilingually to and through their translanguaged reflections.

BTC reflections encompass a range of productive and insightful perspectives about teaching mathematics that are mediated through the dynamic use of bilingual discourse practices. While MALLI practices were explicit areas for discussion, BTC talked most significantly about their identities as bilingual mathematics teachers through the stories they shared in LS meetings that showed a range of interrelated themes and actions (see Table 8.2). BTCs express often underexamined and marginalized intellectual and socio-political work that bilingual education teachers face in schools (Amanti, 2019). Moreover, BTC reflections about and while translanguaging are instances of transcultural repositioning, as cultural work recentering marginalized bodies of knowledge, particularly those that identify tensions and shifts in cultural knowledge related to teaching mathematics bilingually (Guerra, 2015). In conclusion, it is difficult to overstate the significance of creating and sustaining fluid bilingual spaces where BTCs collaborate with experienced bilingual teachers, researchers, and one another.

TRANSCRIPTION CONVENTIONS

Symbol/Feature		Explanation
Bolded Words	Bold	Refers to recorded talk used by speakers
Italicized Words	Italics	Translation of bolded words into English
—	Dash	Abrupt word cut-off or partial utterance
(( ))	Double Parenthesis	Non-Verbal Action/Gestures
“ ”	Quotations	Speech that is quoted in talk or reported speech
,	Comma	Minor pause/less than 1 second
(#)	Parenthesis with number	Number of seconds between utterances
[ ]	Brackets	[inaudible, not transcribable]

REFERENCES

Alfaro, C. (2019). Preparing critically conscious dual-language teachers: Recognizing and interrupting dominant ideologies. *Theory Into Practice, 58*(2), 194–203.

Amanti, C. (2019). The (invisible) work of dual language bilingual education teachers. *Bilingual Research Journal, 42*(4), 455–470.

Bjuland, R., & Mosvold, R. (2015). Lesson study in teacher education: Learning from a challenging case. *Teaching and Teacher Education, 52*, 83–90.

Bravo, M., Mosqueda, E., Solís, J. L., & Stoddart, T. (2014). Possibilities and limits of integrating science & diversity education in preservice elementary teacher preparation. *Journal of Science Teacher Education, 25*(5), 601–619.

Celedón-Pattichis, S., & Ramirez, N. (2012). *Beyond good teaching: Advancing mathematics education for ELLs.* National Council of Teachers of Mathematics.

Cervantes-Soon, C. G., Dorner, L., Palmer, D., Heiman, D., Schwerdtfeger, R., & Choi, J. (2017). Combating inequalities in two-way language immersion programs: Toward critical consciousness in bilingual education spaces. *Review of Research in Education, 41*(1), 403–427.

Cenoz, J., & Gorter, D. (2020). Pedagogical translanguaging: An introduction. *System* (Linköping), 92, 102269. https://doi.org/10.1016/j.system.2020.102269

Druken, B. K., Marzocchi, A. S., & Brye, M. V. (2021). Facilitating collaboration between mathematics methods and content faculty through cross-departmental lesson study. *International Journal for Lesson & Learning Studies, 10*(1), 33–46.

Ek, L. D., Sánchez, P., & Quijada Cerecer, P. D. (2013). Linguistic violence, insecurity, and work: Language ideologies of Latina/o bilingual teacher candidates in Texas. *International Multilingual Research Journal, 7*(3), 197–219.

Flores, B. B., Claeys, L., & Gist, C. D. (2018). *Crafting culturally efficacious teacher preparation and pedagogies.* Lexington Books.

Galindo, R. (1996). Reframing the past in the present: Chicana teacher role identity as a bridging identity. *Education and Urban Society, 29*(1), 85–102.

García, O. (2013). Countering the dual: Transglossia, dynamic bilingualism, and translanguaging in education. In R. Rubdy & L. Alsagoff (Eds.), *The global-local interface and hybridity* (pp. 100–118). Multilingual Matters.

García, O., Flores, N., & Woodley, H. M. (2012). Transgressing monolingualism and bilingual dualities: Translanguaging pedagogies. In A. Yiakoumetti (Ed.), *Harnessing linguistic variation to improve education* (pp. 45–75). Peter Lang.

García, O., & Wei, L. (2014). *Translanguaging and education.* Palgrave Macmillan.

Gort, M., & Sembiante, S. F. (2015). Navigating hybridized language learning spaces through translanguaging pedagogy: Dual language preschool teachers' languaging practices in support of emergent bilingual children's performance of academic discourse. *International Multidisciplinary Research Journal, 9,* 7–25. https://doi.org/10.1080/19313152.2014.981775

Guerra, J. C. (2015). *Language, culture, identity and citizenship in college classrooms and communities.* Routledge.

Guerrero, M. D., & Lachance, J. R. (2018). *The national dual language education teacher preparation standards.* Dual Language Education of New Mexico. https://www.emmastandards.org/_files/ugd/d15f46_55947aa792b34764bd6e62c75a9f95cc.pdf

Langman, J., Solís, J. L., Martin-Corredor, L., Dao, N., & Garza, K. G. (2021). Translanguaging for STEM learning: Exploring tertiary learning contexts. *Translanguaging in Science Education, 27,* 39–60.

Lewis, C., & Perry, R. (2014). Lesson study with mathematical resources: A sustainable model for locally-led teacher professional learning. *Mathematics Teacher Education and Development, 16*(1).

Lewis, C. C., Perry, R. R., & Hurd, J. (2009). Improving mathematics instruction through lesson study: A theoretical model and North American case. *Journal of Mathematics Teacher Education, 12*, 285–304.

Martin, A. D., & Strom, K. J. (2016). Toward a linguistically responsive teacher identity: An empirical review of the literature. *International Multilingual Research Journal, 10*(4), 239–253.

Martínez-Álvarez, P. (2017). Language multiplicity and dynamism: Emergent bilinguals taking ownership of language use in a hybrid curricular space. *International Multilingual Research Journal, 11*(4), 255–276.

Martínez-Álvarez, P. (2020). Essential constructs in the preparation of inclusive bilingual education teachers: Mediation, agency, and collectivity. *Bilingual Research Journal. 43*(3), 304–322. https://doi.org/10.1080/15235882.2020.1802367

Moll, L. C. (2013). *LS Vygotsky and education.* Routledge.

Moschkovich, J. (2006). Using two languages when learning mathematics. *Educational Studies in Mathematics, 64*(2), 121–144. https://doi.org/10.1007/s10649-005-9005-1

Musanti, S. I., & Rodríguez, A. D. (2017). Translanguaging in bilingual teacher preparation: Exploring pre-service bilingual teachers' academic writing. *Bilingual Research Journal, 40*(1), 38–54.

Pierson, A. E., & Grapin, S. E. (2021). A disciplinary perspective on translanguaging. *Bilingual Research Journal, 44*(3), 318–334. https://doi.org/10.1080/15235882.2021.1970657

Solís, J. L., Bravo, M., San Martin, K., & Mosqueda, E. (2018). Preparing preservice teachers to support Latin@ student participation in science practices. In T. T. Yuen, E. Bonner, & M. G. Arreguín-Anderson (Eds.), *(Under)Represented Latin@s in STEM: Increasing participation throughout education and the workplace* (pp. 63–81). Peter Lang Publishing.

Solís, J. L., & Bunch, G. (2016). Responsive approaches for teaching English learners in secondary science classrooms. Foundations of the SSTELLA framework. In E. Lyon, S. Tolbert, J. L. Solís, T. Stoddart, & G. Bunch (Eds.), *Secondary science teaching for English Learners: Developing supportive and responsive learning contexts for sense-making and language development* (pp. 21–48). Rowman & Littlefield Publishers.

Teemant, A., Leland, C., & Berghoff, B. (2014). Development and validation of a measure of critical stance for instructional coaching. *Teaching and Teacher Education, 39*, 136–147.

Tian, Z., & Zhang-Wu, Q. (2022). Preparing pre-service content area teachers through translanguaging. *Journal of Language, Identity & Education, 21*(3), 144–159.

Uribe-Flórez, L. J., Araujo, B., Franzak, M., & Haynes Writer, J. (2014). Mathematics, power, and language: Implications from lived experiences to empower English learners. *Action in Teacher Education, 36*(3), 234–246.

Vomvoridi-Ivanovic, E. (2012). "Estoy acostumbrada hablar Inglés": Latin@ pre-service teachers' struggles to use Spanish in a bilingual afterschool mathematics program. *Journal of Urban Mathematics Education, 5*(2), 87–111.

Wertsch, J. V. (1991). *Voices of the mind: Sociocultural approach to mediated action.* Harvard University Press.

Yeh, C. (2017). Math is more than numbers: Beginning bilingual teachers' mathematics teaching practices and their opportunities to learn. *Journal of Urban Mathematics Education, 10*(2), 106–139.

Zavala, M. (2017). Bilingual pre-service teachers grapple with the academic and social role of language in mathematics discussions. *Issues in Teacher Education, 26*(2), 49–66.

PART III

FAMILY ENGAGEMENT

CHAPTER 9

RECOGNIZING AND EMBRACING PARENTS' RICH MATHEMATICS BACKGROUNDS

Kathleen Stoehr
Santa Clara University

Briana Bravo
Stanford University

ABSTRACT

In this chapter, we explore how the MALLI mathematics workshops for parents shaped the parents' understanding of the role their own mathematics backgrounds play in supporting their child(ren)'s bilingual mathematics education. We utilized a framework that considers parents as intellectual resources, positively viewing their participation and involvement in the mathematics education of their children (Civil & Andrade, 2003) as both participants and sources of knowledge in the mathematics education of their children, regardless of their own formal mathematics background. In our study, the parents attended four workshops that focused on (a) key mathematics content ar-

Mathematics Instruction in Dual Language Classrooms, pages 159–173

eas that included numeracy, geometry, fractions, and open-ended real world mathematics tasks; and (b) the role that language and literacy skills play in learning and understanding mathematics. After attending the workshop series, the parents participated in an individual interview. Key findings from the workshop experience for parents are shared.

A growing body of research indicates the importance of creating strong partnerships in mathematics between home and school (Civil & Andrade, 2003; Civil, 2007; Civil et al., 2019; Colegrove & Krause, 2017; Kelly, 2020; Stoehr & Civil, 2019). Connections that are made between home and school can greatly benefit student learning. This is especially true for mathematics (Anhalt & Rodríguez-Pérez, 2013; Civil, 2007; Civil et al., 2019; Colegrove & Krause, 2017; Kelly, 2020; Stoehr & Civil, 2019). Therefore, providing opportunities that promote authentic and meaningful engagement around mathematics in ways that position parents as active and knowledgeable participants in the education of their children is essential, especially for families from underserved communities (Civil, 2007; Colegrove & Krause, 2017; Kelly, 2020; Stoehr & Civil, 2019).

Civil and Andrade's (2003) work reveals the importance of creating a two-way dialogue between home and school that views parents as *intellectual resources.* To create an authentic two-way dialogue, teachers and schools need to commit to a genuine partnership with parents that is interested and respectful of parents' views and ways that mathematics is used in their homes and communities. Building rapport and trust with families are situated at the core of the parents as an intellectual resources framework. Given that more than 50% of students in public schools are from homes and communities that are culturally, linguistically, and racially diverse, (Kena et al., 2015) it is critical that teachers explore ways to tailor their mathematics instruction to reflect parents' rich mathematical knowledge and experiences. Additionally, Civil and Andrade (2003) also recognize that it is important for parents to learn about how mathematics is being taught at their children's school so that parents can connect school mathematics to the mathematics that takes place in their homes on a daily basis. Such symbiosis between home and school has the potential to optimize student learning and build a healthy sense of identity for them as well.

This chapter examines the interviews of 72 parents at one of the two study sites who upon completing the workshops shared their own mathematics experiences of learning mathematics and their reflections of participating in the workshops. The research was part of a professional development grant titled Math and Language, Literacy Integration in Dual Language Settings (MALLI). While the grant provided professional development to preservice and in-service teachers that focused on three components of instruction in mathematics classrooms (discourse, vocabulary, and literacy), mathematics workshops were offered to parents/caregivers of the students whose teachers who took part in this study. Specifically, five cohorts of parents who had children in MALLI teachers' classrooms received targeted mathematics learning opportunities.

The parents attended four mathematics workshops with the goal of (a) strengthening their own mathematics skills; (b) showcasing their ways of solving mathematics tasks; (c) building their confidence and competence around mathematics; (d) shedding light on the relationship between home mathematical and language practices and the expected learning goals of the classroom; (e) acknowledging and embracing their everyday mathematical interactions with their children; and (f) engaging in mathematics tasks with their children's teachers.

Context

There were 72 parents from the western region of the United States that participated in the workshops. All the parents had at least one elementary aged child whose teacher participated in the larger study. More than 95% of the parents identified as female and Latina. The workshops were offered primarily in Spanish, as most of the parents' first language was Spanish. Each workshop focused on building parents' knowledge of a specific mathematics topic through hands-on activities and the opportunity to work with their child's classroom teacher during one of the workshops. Figure 9.1 showcases the content area of the four workshops.

For example, in the geometry workshop, the different geometric shapes and their attributes and how they align with the common core state standards were discussed and reviewed. Time was also spent exploring how the names of the shapes were similar in Spanish and English (e.g., círculo, circle, rectángulo, rectangle). The parents then built a geometric object that required a minimum of four different shapes. A second requirement was that the object needed to be able to stand up and that the parents identified the shapes in both Spanish and English. Figure 9.2 shows two examples of the objects that were created.

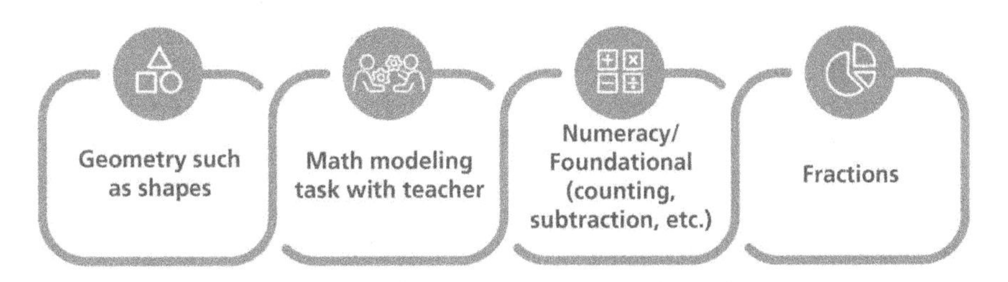

Figure 9.1 Overview of the content areas of the four workshops.

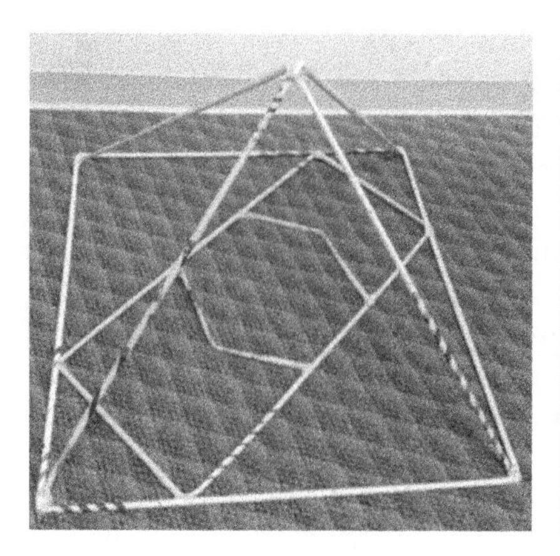

Figure 9.2 Examples of the objects created by the parents.

Additionally, the parents were encouraged to (a) work on a geometry task with their child at home; (b) take a walk in their neighborhood and community to identify geometrical shapes they see; and/or (c) use the geometrical template they were given for participating in the workshop to create an object or creature. The aim was for the parents to utilize their geometric knowledge to support their child's mathematical learning while at the same time supporting their child's mathematics language development in Spanish and English outside of school.

A second example of the mathematics workshops focused on the parents and preservice teachers working together on an authentic and challenging real world mathematics modeling task that showcased the parents' mathematics problem solving skills and understandings. The parents were invited to the preservice teachers' mathematics methods class. The bilingual preservice teachers were paired with their students' parents to work together in small teams to solve a popcorn classroom movie night task (see Appendix for the task). The requirement for the task was that all team members were to contribute to solving the task and present their solution to the class. Figure 9.3 shows two examples of how two parent–teacher teams solved the popcorn task. The parents were then encouraged to create a similar task to solve with their children at home, once again to support their child's mathematical learning while at the same time supporting their child's mathematics language development in Spanish and English outside of school.

At the completion of the four workshops, the parents participated in an individual interview with a MALLI researcher that sought to capture their own experiences with learning mathematics, their descriptions of mathematics that takes place in their daily lives, the role that language plays in learning mathematics, their reflections on having their child in a dual language school, and their reflections on the MALLI mathematics workshops. The

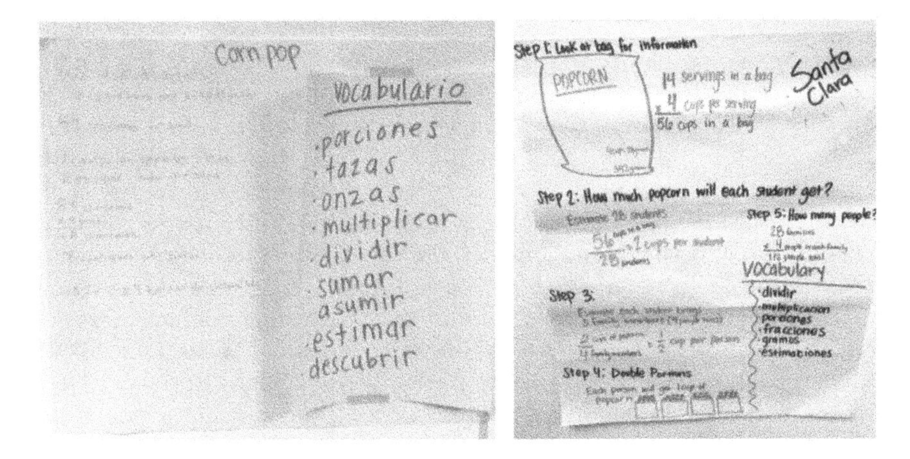

Figure 9.3 Examples of solutions of the popcorn task from two parent–teacher teams.

TABLE 9.1 Sample Interview Questions
How similar or different was your experience to how your child's experience learning mathematics?
What do you remember about the mathematics vocabulary and wording when you were learning mathematics?
Have the workshops helped you see mathematics in a new way? How?
What conversations do you have with your child about mathematics activities that take place in your home? Can you share a specific example?
Do you notice a difference in your child's understanding of mathematics since you were involved in the workshops?
In what ways have the workshops helped you to support your child's mathematics learning?
In what ways have the workshops helped you to learn skills for supporting your child's bilingualism and mathematics knowledge development at home?

interviews were primarily conducted in Spanish, with the option of parents being interviewed in English. Most of the parents chose to be interviewed in Spanish. The interview consisted of seventeen questions that focused on the parents' experiences with mathematics, and reflections that centered on language and the mathematics workshops. Table 9.1 includes a sample of the interview questions. The interviews were approximately 20–35 minutes in length. This chapter now examines the reflections of the parents' interviews.

ANALYSIS

Our first step was to transcribe the interviews in Spanish and then in English for the parents that were interviewed in Spanish as well as transcribe the English interviews. Our first analysis of the interviews focused on thoroughly reading each of the transcribed interviews. We focused specific attention on how the parents responded to each of the seventeen interview questions. Both authors then reread and discussed our initial thoughts of the types of responses that we noticed. We then conducted an iterative analysis (Bogdan & Biklen, 2006) to sort the parents' key ideas across all the interview questions. Examples of these key ideas include the parents' experiences of learning mathematics, the role that language plays in mathematics learning, mathematics in the home and community, supporting their child's mathematics understandings, and learning mathematics in a dual language school. We then wrote analytic memos (Maxwell, 2013) to summarize key ideas across the interviews. We separated the parents' key ideas into five themes that captured the main ideas across the interviews of the 72 parents. These themes are introduced in the findings section.

FINDINGS

The findings in our study examines the interview responses of 72 parents of elementary age children whose first language was primarily Spanish. The interview took place after the parents participated in four mathematics workshops. Table 9.2 reveals the five themes that arose from the parents' individual interviews. Each theme is presented below.

School Learning of Mathematics in the Past and the Present

All 72 of the parents reported that their own experiences of learning mathematics in elementary school was quite different than how their child was learning mathematics at school. For example, one parent spoke of how his daughter was learning to decompose numbers. He stated:

> I think it is more challenging for her. 8 + 4 = 12 but then she needs to circle the ten. It's like dissecting the equation. It's like taking a car apart. It feels like kids are learning pre-algebra. Compared to what I did, it is very different.

TABLE 9.2 Five Themes From the Parent Key Ideas
School Learning of Mathematics in the Past and the Present
Making Connections to Mathematics in Everyday Life
Creating Parent Mathematics Superpowers
Parents' Commitment to Bilingual Education
Math is Fun!

Many parents shared that their children were learning to multiply ways to solve mathematics problems. They believed this was a positive way to learn mathematics, as it provided more ways to arrive at a mathematical solution and to be able to "think a little further" and "put their minds to work a little more." One parent stated that she feels "they [her children] have more opportunities and chances to learn math. Well, they have to pay more attention to the problems that they are studying, so they are thinking and reasoning more."

Another parent said that she was pleased her child was learning multiple ways to solve mathematics problems because of her own experiences of learning mathematics. She reported, "the formulas they taught us were very complicated and they did not teach us the reason why we were doing them that way. The students learned a way, a formula and then finished learning." One parent stated the following:

> My children they are learning that yes there are other ways to do multiplication and other math topics. I did not know that. And also, with my children, they are learning why, why they need those formulas, why they are using formulas, this is different from how I learned; how just remembering the formula and doing it to get the answer.

This parent was pleased that her children were learning mathematics that moved beyond simply memorizing a formula. Instead, they were learning the "whys" of mathematics that afforded them with opportunities to utilize multiple mathematics strategies.

One of the mothers reported her initial discomfort when her child started learning mathematics at school and was being taught different ways to solve a mathematics problem. She said:

> Here there are more ways to get a result, and teachers teach many different ways to do math. At first, I did not like it because I thought it would confuse the kids, but now I think it's good for the kids to do.

These parents' responses highlight the differences in how parents reflected upon how they learned mathematics as elementary aged children and how their children are learning mathematics. The parents were acknowledging that their own experiences learning mathematics involved more passive learning and rote memorization whereas their children were learning to explain their reasoning and were developing a deeper conceptual understanding of mathematics.

Making Connections to Mathematics in Everyday Life

Although the parents all acknowledged that their children were learning mathematics quite differently in school than they had learned this content area, the workshops afforded them the opportunity to reflect and make connections to the different ways that rich and relevant mathematics took place daily in their families' lives. The workshops also served to showcase how much mathematics knowledge each parent possessed as they made multiple connections to mathematics in the real world. Parents shared many examples of distance, time, measurement in cooking, shopping, fractions, and geometry. For example, one parent reported two recent mathematics experiences with her children:

> When I'm in my car and the store is, say about 10 miles from home, and if I am going 65 miles per hour, how long will it take for us to get there? And with the children today we are cooking some cupcakes and I am asking them things with measurements like how much flour do we need, sugar, eggs, and water.

Another parent told of a mathematics time problem she posed to her son:

> It is interesting I was talking with my son this morning. I told him that after school he would be at his grandmother's house from 4:00–8:00. He said that would be 6 hours. I said okay, what time do you get out of school? He said 4:00. I said you will be there until 8:00. How many hours will that be? I had him count from four to eight and he then could see that he would be there for four hours.

One example of a problem involving shopping and money was reported by another parent:

> In shopping when they want an ice cream, I tell them, "How much does it cost?" and tell me, "If I give you a $10 bill or whatever, will you have enough money to pay?" And they convert the amount. They don't get the money until they tell me how much it costs or if they have enough money.

One father talked about how he made a mathematics problem for his children after buying oranges. He said:

> The other day I bought oranges and asked them if they can cut the orange into four parts. I also ask them if we need eight pieces, how could they break the orange? At the same time, I asked them about the shape of an orange and if they could name.

The responses from these parents highlight how the parents were making connections from the mathematics their child was learning in school to the mathematics that took place in their homes on a daily basis. This aligns with Civil et al.'s (2020) work that illustrates the importance of parents being able to make these connections.

Creating Parent Mathematics Superpowers

Many of the parents stated that their own experiences of learning mathematics at school had been difficult and challenging for them. Some parents recalled feeling scared and/or nervous when they were learning mathematics at school. This they shared often left them doubting their own mathematics capabilities. Yet as one parent stated, "Math is needed in everyday life." One parent reflected upon how her experience of learning mathematics at school was one bad experience after another, which often led to her being punished at school and at home.

At the end of the four workshops, many of the parents spoke of how the workshops helped them to see the mathematics knowledge they possessed. One parent reported that the workshops gave him "mathematics superpowers" and that the workshops "have reawakened things for me . . . and got my wheels turning." He also shared that because of the workshops his "awareness in mathematics is sharper." Another parent spoke of how the workshops helped to "refresh my memory." Previous research has documented how parents who feel anxious or less competent in their mathematics abilities can gain mathematical confidence when provided with opportunities to reexamine their mathematics abilities (Civil et al., 2020).

Other parents stated that attending the workshops helped them to revise and/or relearn mathematics in a safe and supportive space. As one mother said, "Today I told my son that I, too, am learning math." One mother shared how she felt when she attended the first workshop. She stated, "At first, I was embarrassed to say that I hadn't understood [the mathematics content]. Then I said to myself, 'I'm here to take advantage of being at the workshops.'"

Another parent reported that as a result of attending the workshops, she could help her children more with their mathematics homework. She added, "Before this year I did not know where to start. Today I have an idea on how to start and some skills on how to solve those [mathematics] problems." These responses illustrate how creating mathematics workshops for parents can create a safe space for parents to "reawaken, "refresh'" and "learn mathematics."

Parents' Commitment to Bilingual Education

The parents shared similar reasons for why they chose a dual language school for their child. Many parents reported that Spanish was the language spoken in their homes and by their extended family members. One parent talked about the importance of their children being able to communicate with extended family members who spoke only Spanish. Another parent stated,

> First of all, my language is Spanish, so they are going to learn Spanish because it is the language we speak at home. We chose this school because we want our children to be strong in both languages, Spanish and English.

Similarly, one parent shared the importance of their child "not losing their language." She added, I do not speak English. I have seen cases of parents who do not speak English and their children do not speak Spanish and cannot communicate. My husband and I wanted them not to lose their language and for that reason we chose a school that had dual immersion.

Other parents talked about the relationship between Spanish and culture. One parent reported, "My children need to maintain their culture because language is part of culture." Some parents spoke about how knowing Spanish and English could be useful to them in their everyday lives as well in the future. One mother stated that "bilingualism is good because they can help someone on the street who can't speak English like me."

Some parents believed that being bilingual in Spanish and English could lead to their child being able to get a better job in the future. One mother said, "I think that more doors are opened for them so that they can develop or be better in the future because there are more job opportunities for someone who speaks both languages." Another mother also agreed that being bilingual could create better job opportunities for her children. She shared the following dream for her daughters:

> My goal for my daughters is to be someone in life. Which is something they always hear from me. I don't want to see them like me working. I clean houses and the work is heavy. I tell them, I don't want that for you, I want you to be professionals and make you want to study. This is my dream, to see them graduate.

One parent spoke of how being bilingual not only could "open doors" for her children but also be of service to other people. She stated:

> Being bilingual opens more doors for you because this country is made up of pure immigrants from Africa, Asia and from different parts. The environment that you see here in California is that the majority are immigrants who speak Spanish. I see that there is a lot of Spanish and I tell him, "How nice, because they can help people who speak Spanish and translate into English or people from English to translate into Spanish." It is a good service to give to others and it can also give you more job opportunities. Help more people. You know I see that they are so lucky to have both languages and I say, "Take advantage of it."

A common theme that some parents shared was that dual language schools afforded them with the opportunity to communicate with their child's teacher in Spanish or English. One mother reported, "I like that their teachers and I can communicate in Spanish, and they can also guide me...For me it is important that parents and teachers can communicate because that way we know what our children need to get ahead." Another mother stated that "being able to communicate with the teacher in Spanish on ways to improve and help her child in mathematics was key." This, she said, made her feel more comfortable in asking for support from the teacher. Lastly, one parent said that learning mathematics in both Spanish and English makes "mathematics even better." The responses from these parents emphasize the power and beauty that bilingual education can create for parents and families whose home language is Spanish. Previous research collaborates the importance of providing parents whose first language is Spanish with opportunities to engage with their children's teachers in their home language around mathematics (Colegrove & Krause, 2017; Stoehr et al., 2022).

Math Is Fun!

Many of the parents spoke of how the workshops provided them with the opportunity to see that "math is fun." For example, one parent stated that the workshops helped her to see that many strategies could be used to arrive at a correct answer. She reported, "I did not think there are so many ways that you can do math and in so many fun ways." Some parents stated that they learned that working together with other parents made math more enjoyable and fun as they [the parents] laughed and solved problems together.

Another parent reported that while participating in the fractions workshop, a topic that she felt unsure about, she discovered that "I did not know it was possible to learn math and have fun." One parent spoke about how the manipulatives she received during the fractions workshop resulted in her having fun while learning fractions. She reported, "The teacher brought us fractions in plastic squares and gave them to us. This helped me, and I learned it is fun not only to visualize the problem, but also to touch and see the examples." Our study supports earlier research that documents the importance of ethnically and linguistically diverse parents viewing mathematics as fun (Civil et al., 2002).

Many parents also revealed that because of their participation in the workshops they learned ways to support their child's mathematical understanding while making mathematics fun for them. One parent shared the following reflection:

> Sometimes math is hard for my child, he or I would get frustrated, and in the end I couldn't even help him. Now that I have taken the workshops, I realized

> that he can learn in an easy and fun way like playing card games like play split. That was what I liked a lot, we put it [mathematics] in the form of a game and my child understands it better.

Another parent also spoke of how the workshops helped her to see mathematics as more "like a fun game" that she could do with her children. She added that the games she learned also provided her with the opportunity to "show our children that we can also do math by playing games with them. One parent also shared that because of the workshops, she tries to explain mathematics in fun ways to her daughter so that her daughter can enjoy and retain the mathematics information she is learning.

One mother reported that the workshops helped her to teach her young son that mathematics is fun. She shared:

> We can count little rocks and beans and he tells me, "Oh yeah." There are times when I make him drawings of cereal or little beans or whatever it may be to see how many he counts, and he says that it is fun and very enjoyable.

DISCUSSION

The interviews of these 72 parents offer insight into the benefits of providing parents with bilingual workshops in English and Spanish. Opportunities to work with their child's teachers, strengthened their numeracy skills, and boosted their confidence in mathematics. The results from the interviews also demonstrate that parents can play an essential role in the development of their children's mathematics skills and bilingual proficiency. Ultimately, the four bilingual mathematics workshops taught parents new ways to solve mathematical problems, increasing their confidence to provide support to their child's mathematical learning.

Parent confidence is also an important construct to consider from the data collected for this study. Parent's confidence with mathematics increased in part due to the fact that the workshops were conducted in Spanish. Our study supports Civil et al.'s (2023) work that is grounded in the belief that in order for parents whose first language is Spanish, to have successful mathematics learning experiences, these experiences must be offered in their home language. This strong confidence is a critical precursor for parents to be able to in turn help support their children's mathematical knowledge, either through homework that is sent home, or the mathematics needed to function in their community. This type of collaboration between parents and teachers can help bilingual children build the type of knowledge that builds mathematics success. This begins with a recognition that, as one parent stated, "We the parents are the first teachers for our children."

LIMITATIONS

There are some limitations of our study that we would like to address. First, we rely on the individual parent interviews to understand how the parents processed their experience of participating in the mathematics workshops. While conducting individual interviews is an appropriate way to report on the workshop experiences for the parents, we acknowledge that our study could have been strengthened with researcher observations and field notes of the parents engaging in the mathematics workshops. Additionally, focus group interviews with the parents could also have added another dimension to this study. Second, our study is limited in that our work does not follow up with the parents over time, investigating how and if the workshops had a continued impact on the targeted mathematics learning opportunities (see p. 161 for the six targeted goals) for the parents.

In the future, we hope to expand this work to include additional opportunities for parents to engage in more mathematics workshops together around topics that they are interested in and that they believe would be helpful in supporting both them and their children.

CONCLUSION

Our study offers insights into how offering mathematics workshops to parents whose first language is Spanish have the potential to strengthen parents' mathematics skills, honor and showcase their ways of solving mathematics, build their confidence and competence around mathematics, acknowledge and embrace the rich mathematics that occurs in their home and community on a daily basis, and reinforce their commitment to a bilingual education for their children. Our study supports previous work that reveals the importance of providing opportunities that promote authentic and meaningful engagement around mathematics in ways that position parents from underserved communities as active and knowledgeable participants in the education of their children (Civil, 2007; Colegrove & Krause, 2017; Kelley, 2020; Stoehr & Civil, 2019). We argue that providing these types of learning experiences for parents may strengthen the relationship between home and school and ultimately have a positive impact on students' mathematics learning.

APPENDIX
The Popcorn Modeling Task

Source: Adapted from mathematical modeling with cultural and community contexts (Turner et al., https://sites.google.com/qc.cuny.edu/m2c3/). The task: Popcorn for class movie.

Your class is having a movie night for students and their families. You will serve popcorn as a snack. You will get giant bags of popcorn to share and scoop out servings for each person.

How many giant bags of popcorn does your class need for the movie night?

You need to make sure there is enough popcorn, but not a lot of left-over popcorn.

Make a plan for how to share the popcorn.

Your plan to must show how:

- everyone gets enough popcorn
- it is a fair plan
- you can use the plan in other sharing situations

You can use pictures, numbers and words in your plan.

Questions to think about:

- What do you know?
- What do you need to find out?
- What do you need to *assume*?

REFERENCES

Anhalt, C. O., & Rodríguez-Pérez, M. E. (2013). K–8 teachers' concerns about teaching Latino/a students. *Journal of Urban Mathematics Education, 6*(2), 42–61.

Bogdan, R., & Biklen, S. K. (2006). *Qualitative research for education: An introduction to theory and methods* (5th ed.). Pearson.

Civil, M. (2007). Building on community knowledge: An avenue to equity in mathematics education. In N. S. Nasir & P. Cobb (Eds.), *Improving access to mathematics: Diversity and equity in the classroom.* Teachers College Press.

Civil, M., Guevara, C., & Allexsaht-Snider, M. (2002). Mathematics for parents: Facilitating parent's and children's understanding in mathematics. *Proceedings of the twenty-fourth annual meeting of the North American Chapter of the International Group for the Psychology of Mathematics Education,* 1755–1765. https://www.researchgate.net/publication/316657035_Mathematics_for_Parents_Facilitating_Parent's_and_Children's_Understanding_in_Mathematics

Civil, M., & Andrade, R. (2003). Collaborative practice with parents: The role of the researcher as mediator. In A. Peter-Koop, V. Santos-Wagner, C. Breen, & A. Begg (Eds.), *Collaboration in teacher education: Examples from the context of mathematics education* (pp. 153–168). Kluwer.

Civil, M., Stoehr, K. J., & Salazar, F. (2019). Learning with and from immigrant mothers: Implications for adult numeracy. *ZDM Mathematics Education 52*(3), 489–500. https://doi.org/10.1007/s11858-019-01076-2

Civil, M., Salazar, F., Hernandez, M., Mata, S., & Murdock, L. (2023). Mothers as co-facilitators of mathematics learning experiences. In M. Caspe & R. Hernandez (Eds.), *Family and community partnerships: Promising practices for teachers and teacher educators* (pp. 111–118). Information Age Publishers.

Colegrove, K., & Krause, G. (2017). "Lo hacen tan complicado": Bridging the perspectives and expectations of mathematics instruction of Latino immigrant parents. *Bilingual Research Journal, 40*(2), 187–204. https://doi.org/10.1080/15235882.2017.1310679

Kelley, T. (2020). Examining pre-service elementary mathematics teacher perceptions of parent engagement through a funds of knowledge lens. *Teaching and Teacher Education, 91,* 1–14.

Kena, G., Musu-Gillette, L., Robinson, J., Wang, X., Rathbun, A., Zhang, J., Wilkinson-Flicker, S., Barmer, A., & Dunlop Velez, E. (2015). *The condition of education 2015* (NCES 2015-144). U.S. Department of Education, National Center for Education Statistics. http://nces.ed.gov/pubsearch

Maxwell, J. A. (2013). *Qualitative research design: An interactive approach* (3rd ed.). Sage.

Stoehr, K., & Civil, M. (2019). Conversations between preservice teachers and Latina mothers: An avenue to transformative mathematics teaching. *Journal of Latinos and Education,* 1–13. https://doi.org/10.1080/15348431.2019.1653300

Stoehr, K., Salazar, F., & Civil, M. (2022). The power of parents and teachers engaging in a mathematics bilingual collaboration. *Teachers College Record, 124*(5), 30–48. https://doi.org/10.1177/01614681221103947

CHAPTER 10

'TO KEEP NUESTRA CULTURA'

Math, Language, and the Importance of Bilingual Spaces for Bilingual and Latinx Parents

Brenda Sarmiento-Quezada
Purdue University

ABSTRACT

This chapter explores parent involvement and participation in three bilingual mathematics workshops offered in Title 1 schools in central Texas. Specifically, I studied Latinx bilingual parents' perceptions of how language supports their child's bilingualism and how their participation in these workshops increased their own understanding of mathematics. Using interview data from The Mathematics and Language, Literacy Integration (MALLI) in dual language settings project, this chapter highlights the voices of Latinx parents and the cultural importance they place on having their children in a dual language program. This chapter acknowledges the need to create workshops designed specifically for Latinx bilingual parents to better bridge communication between school and home practices.

Mathematics Instruction in Dual Language Classrooms, pages 175–190

Research has shown that parents and family members who interact with children during the early years of life become integral contributors to children's initial experiences with the development of literacy skills (Dearing et al., 2006; Englund et al., 2004; Hart & Risley, 1995; Nievar et al., 2011). From birth, children begin imitating behaviors that contribute to literacy as they follow the functions of conversations, writing, and reading (Lave & Wenger, 1991). Research promoting and supporting family engagement in literacy, then, is well documented (Dearing et al., 2006; Englund et al., 2004; Hart & Risley, 1995; Nievar et al., 2011). In mathematics, even as research examining mathematics development and parent involvement has slowly emerged in the last decade (Chang et al., 2015; Civil & Quintos, 2022; Colegrove & Krause, 2017; Stoehr et al., 2022; Wen et al., 2012; Wilder, 2017); the role and importance of families in mathematics education, particularly in bilingual contexts, continue to be understudied. Parents who practice language and mathematics education at home may not be aware of their contribution to their children's mathematics education in their daily activities. Therefore, this research focuses on examining the role of parental and family engagement in mathematics through parent workshops, particularly within culturally and linguistically diverse families whose children are emergent-bilinguals (EB).

Research has indicated the importance of strong early-childhood mathematics education for later learning (Duncan et al., 2007; Watts et al., 2014). While schools certainly play a role in strengthening early childhood mathematics education, there are multiple opportunities outside school, namely in the family setting that are mathematically rich but may go unnoticed by parents and teachers. Current findings also show that although parents support their children's school success, they often struggle with finding ways to help their children with mathematical learning, as compared to literacy (Cannon & Ginsburg, 2008; Segers et al., 2015). Moreover, research on early childhood education highlights the need for connecting home practices and children's mathematical content for learning achievement in bilingual contexts (DeFlorio & Beliakoff, 2015; Kleemans et al., 2012; Moll et al., 1992).

For parents, supporting mathematics at home poses uncertainty on how to help their children or whether they should be trying to do so (Cannon & Ginsburg, 2008). The literature on early childhood education and development points to the need to develop mathematical knowledge in parents and families so they feel better positioned to develop mathematics learning opportunities for their children (Cannon & Ginsburg, 2008; Kleemans, & Varhoeven, 2015; Nievar et al., 2011). The most essential aspect for us to consider while conducting the MALLI project was to view parents as intellectual resources (Civil, 2016), in order to facilitate an authentic two-way dialogue between school and home practices. To do so, schools must genuinely be interested in parent and family views, their use of mathematics at home, and explore ways in mathematics instruction that reflect parent and family knowledge and experiences. At the same time, the families of EB students can find benefits to learning about the school approaches to teaching mathematics and begin to explore connections between school mathematics and their everyday interactions with their children.

METHODS

This chapter draws from the parent and family involvement component of MALLI, a larger five-year research project examining bilingual pedagogy in mathematics, language, and literacy integration in California and Texas. Particularly, this chapter centers on promoting family engagement in mathematics in the dual-language setting. During three school cycles, participating parents attended four mathematics workshops that presented a variety of activities to activate mathematical knowledge, integrate the home-linguistic practices, and connect them with school teaching to support the mathematics education of EB students. The activities in these workshops focused on three mathematics areas: discourse, vocabulary, and biliteracy.

These workshops took place during the 2019, 2020, and 2021 Spring semesters in three Title 1 school districts located in Texas and California. This chapter focuses only in the Texas context where participating parents came from different academic and professional backgrounds and whose children belonged to Spanish-English two-way dual language classrooms. Since this study draws from a larger research project, parent recruitment was done only in the classrooms of bilingual teacher candidates (BTC) and bilingual mentor participants that were a part of the study. An interest and availability survey was developed and administered to parents by the research team. At the end of the three cycles, 22 parents participated in our workshop series in Texas. It is important to note that during the 2019 cycle, all sessions were in person. In 2020, two sessions took place in person and the remaining two were joined virtually by parents, BTCs, and the research group as emergency protocols were enacted in response to COVID-19. The last cycle, 2021, remained virtual since school districts in Texas were still operating under strict health guidelines.

Each workshop had a specific mathematical focus and a language learning goal. The focus and goals were identified as ones that BTCs expressed difficulty teaching. Workshop 1 and 3 centered around geometric and fraction tasks, respectively. These workshops presented parents ways on how to intentionally and purposefully use everyday vocabulary and domain-specific vocabulary to reinforce mathematics with activities at home, such as cooking or through play. Workshop 2 presented several numeracy tasks involving decimals and allowed parents to closely read literary and informational texts to determine how meaning is conveyed explicitly and implicitly through language. Lastly, workshop 4 was joined by BTCs of the larger research project. Parents and BTCs collaborated to create a mini-lesson that could take place at home on measuring and estimating length. This workshop aimed at providing a learning experience for both groups on teacher–parent partnership and collaborative problem-solving related to academic topics. All four workshops involved activities that could be implemented at home, gave parents different opportunities to use mathematical vocabulary in daily conversations, and engaged them in bilingual reading and writing activities with a mathematics focus. The emphasis of these workshops was on K–3, but provided some foundation for the upper elementary grades.

In addition to these workshops, parents completed a pre and post 7-item family survey on their and their children's perceptions about mathematics (as reported by the parents), along with a parent profile in which they indicated their level of education and level of confidence with helping their children in mathematics. Moreover, these parents participated in a one-time in-person interview at the end of the four workshops. During the interviews, parents were asked to compare their own past experiences with mathematics to the ones their children are currently experiencing, along with questions related to activities they do at home to enhance mathematical and literacy practices with their children.

DATA SOURCES

While the study took place in California and Texas, this chapter focuses on survey and interview data collected at the end of the series of workshops in Texas. This chapter looks at the responses of 22 parents in Texas that participated in the workshops during the 2018–2019, 2019–2020, and 2020–2021 academic years. All participating parents self-identified as female and came from a range of educational backgrounds. For instance, about 46% of parents held a high school diploma or GED, whereas 23% had a technical or associate degree. Additionally, 18% of parents graduated college with a bachelor's degree, while 13% held a graduate degree, including one parent with a PhD. Eighty-five percent of participating parents were born outside the United States, mostly from México and one from Guatemala, and the remaining were born in the United States. Twenty-one parents self-identified as Hispanic and one as Caucasian.

During the initial survey, the majority of parents indicated their ability to understand, speak, read, and write English and Spanish. Therefore, although the workshops were designed to be conducted in Spanish, English was used as needed and for parents whose home language was not Spanish.

Attendance at the workshops fluctuated due to varying parent schedules; however, the majority of parents attended 75% or more of the workshops.

Interviews took place after the last workshop, and they ranged from 20 min.–45 min. These interviews were conducted in person or virtually, at a convenience of the parent, and in their preferred language. Interviews consisted of seventeen questions and were divided in three domains: parent experiences with mathematics, language reflections, and workshop reflections. In addition, pre- and post- surveys were given at the beginning of Workshop 1 and at the end of Workshop 4. These surveys had the option to be completed anonymously; however, parents opted to identify themselves while completing them.

ANALYSIS AND RESULTS

This study adopted a reflexive thematic analysis approach, adhering to an inductive methodology as proposed by Clarke and Braun (2013). Parent interviews were conducted with the participants' consent, and their responses were transcribed. A total of 22 audio-recorded interviews, each ranging in duration from 20 to 45 minutes, were analyzed. Within the context of these interviews, participants shared their experiences concerning mathematics, activities conducted within their households, and their involvement in mathematics workshops.

The primary phase of analysis involved an exhaustive examination of each transcribed interview, with particular attention directed towards parental responses to the interview queries. Subsequently, I re-read and engaged in conversations with colleagues of this project with the goal of clarifying noticeable patterns in responses. This iterative process ultimately facilitated the emergence of salient themes across the three cohorts of participating parents.

In my initial analysis, I identified recurring themes, particularly common in bilingual settings, such as language and cultural challenges. Consequently, I conducted a more focused examination where I focused on instances where parents expressed ways in which they perceived mathematics education, either through their own experiences growing up, or as parents.

In this subsequent analytical phase, our goal was to comprehend both how and what parents conveyed through these reflective interviews. This enabled us to identify and substantiate three additional themes derived from their experiences and relationship with mathematics, either through their recollections as students or as parents. These themes encompassed a range of aspects and allowed us to observe instances where parents expressed their perceptions on mathematics, the way they engaged at home with it, and how these workshops empowered them as parents.

Subsequently, the pre- and post-surveys were collected and analyzed by an educational consulting company, which thereafter presented their findings to the entire team. The surveys in question were designed to solicit feedback from parent participants concerning their self-assessed proficiency in assisting their children with mathematics-related activities and home-based practices.

Pre- and Post-Survey

These surveys asked parent participants about their perceived ability to support their child in mathematics activities and home practices. Survey results showed that parents generally engaged in helping their child with homework in reading and mathematics. Nevertheless, they felt most confident when helping their child with reading. The initial survey also indicated 61% of parents found it difficult to help their children with mathematics homework, however, the follow up survey (after the workshops) reflected a decrease, indicating that 39% of parents found it difficult.

Differences were found when comparing responses between parents who attended college and those with less formal education (high school diploma or less). Among parents who did not attend college, 56% said they have the mathematics skills to help their child be successful in mathematics class compared to 91% of parents with at least some college.

One-hundred percent of parents reported that participating in the workshops improved their ability to support their children's bilingual development. At the end of the program, 96% of parents indicated having the mathematics skills to support their children's mathematics learning, which showed an increase of 30% above the baseline rate. Parents also overwhelmingly reported that they benefited from the workshops. Specifically, after each workshop all or almost all parents reported learning something they could share with their child, having more confidence in supporting their child's mathematics and literacy, having more confidence in their own mathematics knowledge, having more confidence in volunteering at school, and intending to support their child's learning at home.

Interviews

This section delves into a comprehensive review of the transcribed interviews, with a particular emphasis on parental responses, leading to the recognition of noteworthy response patterns and the emergence of significant themes among the three participant cohorts. The subsequent analysis phase further explores four recurring themes, particularly in bilingual contexts, and conducts a detailed examination of parents' perspectives on mathematics education based on their experiences and roles, ultimately shedding light on their perceptions, home involvement, and the influence of workshops on their parenting.

Parental Engagement and Cross-Cultural Challenges in Mathematics Education

Our interview analysis indicates that parents appreciated the opportunity to actively participate in the workshops. They also found it helpful to learn vocabulary and hands-on activities they could use at home with their children. Moreover, our analysis indicates parents perceived their own experiences with mathematics significantly different from their children, and some believed this difference interfered with their ability to help with their homework. For instance, as parents were reflecting upon their own experiences with mathematics, they believed the way their children learn today is more complicated than the way they learned themselves. As one parent reflected,

> I'm surprised my 5 year-old, he knows way more math concepts that I knew when I was his age... I'm amazed every time I see my 5 year old counting by 5s by 2s, counting backwards from 20 to 0... those are complex math problems for a 5 year old. I remember my kindergarten year was like having fun, I don't remember having homework back then. So, it's completely different... in a good way.

Also, parents who were born and received their elementary education in México or Central America, expressed some confusion about the methods employed here in the United States. As indicated during their interviews, some of these participants faced challenges when attempting to instruct their children using their own problem-solving methods, as their children would assert that the methods taught in class diverged from their parents' approach or that their parents lacked the knowledge or understanding of problem-solving, as it did not conform to what they had learned in the classroom.

> *Hay muchas de las veces que quiero enseñarles o decirles, se enojan y me dicen "no tú no sabes, asi no me las enseñan a nosotros" "así no es" "déjamelo hacerlo yo sola" y pues te vuelvo a decir el proceso, los procesos de cómo se enseñan las matemáticas, cómo se enseñan aquí, de cómo se enseñan en México, sí son diferentes.*

> There are many times I want to teach them and tell them, but they get mad at me and they tell me, "No, you don't know, this is not how they teach them to us"; "It's not like that"; "Let me do it myself"; and so I tell you, it is the process, the process of how mathematics is taught here and how is taught in México is very different.

These parents also demonstrated concerns about having a negative impact if they share their methods with their children, so they tend to rely on other relatives to help them with mathematics.

Mathematical Engagement Through Everyday Activities

Our findings also showed that some of the most common mathematics activities that parents do with their children at home involve time, cooking, and money. Some of these activities include accounting for video game and travel times, baking and dinner preparation, and savings to buy new toys. One of the participants, for example, mentioned how her daughter enjoyed baking and used this as a way to incorporate mathematics at home. Moreover, she added she would play multiplication games with her daughter in order for her to practice and develop her mathematics knowledge during family trips.

> *Bueno una es cocinar porque a ella le gusta hacer postres entonces con las medidas. La otra es cuando viajamos, le empiezo a decir la multiplicación, porque yo siempre he estado ahí con mi niña tratando de que avance.*
>
> Well one is cooking because she likes making desserts with the measurements. The other one is when we travel, I start asking her the multiplication tables, because I have always been there for my girl to get her moving forward.

Even amidst the COVID-19 pandemic, parents actively participated in virtual sessions and seized the chance to engage in various activities and games with their children, adhering to the state's stay-at-home protocols. One of the parents shared Figures 10.1 and 10.2 with the researchers after Workshop 3, illustrating their children's joint baking experience and involvement with measurements and fractions. In the subsequent weeks, as schools continued to operate remotely, this parent consistently updated the researchers on how their children persisted in practicing mathematical concepts through cooking and baking at home.

Language Challenges and Bilingual Identity in Mathematics Education

Our analysis also found that parents sometimes perceived their own linguistic repertoire as a challenge to be able to help their children with Mathematical concepts. Even as parents in a dual-language program where the language of instruction was in Spanish, when it came to Spanish-translated textbooks they faced difficulties with understanding the texts in Spanish. One of the parents explained,

> *A veces también la traducción de los libros es diferente y no logro entenderla yo, entonces yo tengo que decir "que está tratando de decir aquí?"*

Sometimes the translation of books is also different, so I have to ask myself, "What are they trying to say here?"

Moreover, parents also understood some of the challenges came with having learned the concepts in Spanish themselves, whereas their kids were learning them in English, speaking about this challenge, one parent

Figure 10.1

Figure 10.2

commented: "*Como mis hijos son bilingües, a veces se saben el concepto en inglés y yo en español, entonces a veces batallamos*" / "Since my children are bilingual, there are times when they know the concept in English and I know it in Spanish, so sometimes we struggle." Yet, even with linguistic struggles, parents recognized and placed at the forefront the importance of nourishing and maintaining their children's bilingualism and how linked it was to their own identity. One U.S.-born parent who self-identified as Latina expressed,

> *Me interesa que los niños aprendan más de un idioma, pero sobre todo que no olviden, que siempre tengan presentes sus raíces. Nosotros somos americanos porque nacimos en Estados Unidos, sí, pero nosotros somos mexicanos entonces son méxico-americanos; para que conozcan sus tradiciones y raíces, conozcan nuestros orígines.*
>
> I am interested in children learning more than one language, but above all, I want them to not forget and always remember their roots. We are Americans because we were born in the United States, yes, but we are Mexicans, so they are Mexican-Americans. I want them to know their traditions and roots, to know our origins.

Over the course of the 3-year duration of these workshops, it became increasingly apparent each year that parents highly valued the cultural significance of enrolling their children in a dual language program, "That is something that's been really important to us as a family, when we started the family to keep *nuestra cultura* and being bilingual is a huge part of that, being bicultural is a big part." Some expressed never having the opportunity to learn another language during their schooling, so giving their children this opportunity would give them the ability to connect to their culture and roots, as one parent mentioned, "Because I wasn't able to learn Spanish when I was younger, so I wanted to be different for them, so they were able to learn both languages, and they would be able to communicate with grandparents." And to some of them, having the ability to enroll their children in a bilingual program was the culmination of their own sacrifice and struggles as immigrants themselves. When asked about why it was important for her children to speak both languages, one parent responded:

> I keep telling them, "mijo and mija, I was able to pick [English] up just by listening to people," I came to the United States when I was 20 years old. *Yo llegué ya grande, yo llegué directo a trabajar* [I arrived when I was already grown up, I arrived directly to work]. We wanted to give the opportunity to our children to have not only the ability to speak the language, but also to read it and write it.

Furthermore, all 22 participants in our workshops conveyed their commitment to nurturing their children's bilingualism, and they recognized it as a form of linguistic capital that would enhance their children's employment prospects and contribute to building a more promising future, *"Que sean más valorados en cualquier trabajo que tengan y hasta ganar más dinero*" / "So that they are more valued in any job they have and even earn more money." For all participating parents, then, being fully bilingual could provide more opportunities in the work field to their children and as one parent expressed, her efforts to foster her children's bilingualism was a small way she could help provide the best that this country has to offer, *"Quisiera darle un poquito de este país que tiene lo mejor."*

EMPOWERING PARENTS AS PARTNERS IN MATHEMATICS EDUCATION

Aside from the cultural and linguistic importance of having their children in a dual language program, and supporting the post-survey results, all participants found the workshops helpful in seeing mathematics in a new way and as part of their every-day lives. Parents particularly enjoyed that activities involved games and allowed for group work, something they do not recall having experienced in their own schooling. As one parent expressed,

> *Lo que me gustó mucho de los talleres fue que hicimos muchas actividades en grupo para aprender las matemáticas y el vocabulario de matemáticas en español. En matemáticas, en general, fue algo que yo no me acuerdo haber aprendido de niña, y se me hace muy importante para que los niños puedan ver que las matemáticas pueden ser muy divertidas, se pueden ayudar unos a los otros, en los juegos.*
>
> What I really liked about the workshops was that we did many group activities to learn mathematics and mathematics vocabulary in Spanish. In mathematics, overall, it was something I don't recall learning as a child, and I find it very important for children to see that mathematics can be fun, that they can help each other in games.

Furthermore, and as shown in the post-survey analysis, parents expressed confidence and demonstrated excitement about learning something they could share with their children at home. Moreover, the parents engaged in these workshops were appreciative for having found a community that comprised fellow bilingual parents, enabling them to recognize they were not isolated and that their own-perceived struggles with language and mathematics were shared experiences among parents.

> *Me pareció muy interesante el curso, muy divertido, aprendimos, me gustó mucho trabajar en equipo porque aprendí de las experiencias de otras mamás. No se me hubiera ocurrido hacerlo como lo hacía la otra mamá, se me hizo muy interesante. Ojalá hubiera cursos así para los papás, porque hay veces que no se involucran todos en la forma educativa de los hijos, porque no existe un curso así para que también aprendamos cómo ayudarle a sus hijos.*

> I found the course very interesting, quite enjoyable, and we learned a lot. I particularly enjoyed working as a team because I learned from the experiences of other moms. I wouldn't have thought of doing things the way another mom did, and I found it very fascinating. I hope there are courses like this for dads too, because sometimes not everyone gets involved in their children's education. It's because there isn't a course like this for us to learn how to help our children.

One of the participants, as seen in the above excerpt, articulated that the workshops were not only enjoyable and engaging but also provided an opportunity to collaborate with other mothers, enabling her to benefit from their shared experiences. She emphasized that such interactions allowed her to learn alternative problem-solving approaches in mathematics. This participant expressed a desire for more workshops of this nature to be available to parents, particularly because it can be challenging for parents like her to actively engage in their children's education without the guidance or space provided by such workshops.

Upon conducting and analyzing parent interviews, it also became apparent that for certain parents, active involvement in their children's education was synonymous with effective parenting. In essence, assisting their children with school tasks, especially in the realm of mathematics, was perceived as fulfilling their parental responsibilities. One participating mother, for instance, not only expressed gratitude for the time and guidance provided to parents like her during the workshops, but also expressed a desire for additional workshops in the future to continue her personal growth as a mother, as well as for the collective growth of all mother participants.

> *Estoy muy agradecida nada más por el tiempo que se nos brindó, la capacitación estuvo muy bien, ojalá haya más talleres como estos, que podamos seguir creciendo como mamá, que podamos aprender a ser mejores madres.*

> I am very grateful, primarily for the time that was provided to us. The training was very good. I hope there are more workshops like these, so we can continue to grow as mothers and learn to be better moms.

In addition, when asked about their own mathematics understanding, another of the participants added,

> We learn math a certain way, but that doesn't mean it is the only way. Sometimes we as a parent, we are very closed minded, sometimes we are like "this is the way I learned, this is the way it should be," and that is not true . . . Como dicen en español, *el órden de los factores no altera el producto,* learning that there is one more way to solve a problem, it was challenging at the beginning, I learned a lot from workshops like these one that there is more than one way to learn one concept, how to solve a problem, and it's really amazing.

In this quote, the parent reflects on the way she was taught mathematics and how sometimes parents have a closed-minded perspective, believing that the way they learned is the only correct way and acknowledging that this viewpoint is not accurate. She code-switches and uses the phrase *"el órden de los factores no altera el producto"* (in Spanish) underscoring the idea that there can be multiple approaches to solving a problem or understanding a concept in mathematics. This parent expresses that initially, it was challenging for her to accept this idea, but through these workshops and similar experiences, she learned that there are multiple valid ways to approach learning a concept or solving problems in mathematics, emphasizing the importance of being open to different approaches to learning and problem-solving.

Thus, and as reflected in the post-survey results and in the interviews, parents who participated felt more comfortable in their own mathematics knowledge and expressed a better understanding of the role of literacy in mathematics. The workshops provided by this project underscore the significance of establishing inclusive environments for bilingual and Latinx parents. These spaces serve as platforms for parents to not only enhance their own language and mathematics proficiency but also foster connections with fellow parents, enabling the formation of a supportive network. This network equips parents with the essential resources to effectively support their children's education at home, thereby fostering improved communication and bridging the gap between school and home practices.

DISCUSSION AND SCHOLARLY SIGNIFICANCE OF THE STUDY

The primary objective of this study was to explore and gain a deeper understanding of the role and knowledge that parents and families bring to mathematics instruction. The findings of this study provide evidence of the positive impact that educational practices can have when they involve parents in a reciprocal dialogue regarding their children's mathematical education (Stoehr & Patel, 2018). Notably, the study revealed that parents are willing to learn and apply school practices in their daily lives when provided with the opportunity. As such, this research underscores the significance of engaging and empowering parents as active partners in their children's education, paving the way for improved educational outcomes and enhanced connections between home and school.

Parents and family members have a critical role in children's education in their early years. They are an essential part in their literacy skills development (Dearing et al., 2004; Hart & Risley, 1995; Nievar et al., 2011). This chapter acknowledges the need to create workshops designed specifically for Latinx bilingual parents to effectively establish a connection between school and home practices. Moreover, the objective of this chapter is to challenge and rectify misconceptions surrounding Latinx parent engagement, care, and comprehension regarding their children's education, with a specific focus on mathematics education.

The incorporation of bilingual and Latinx parents in education holds significant importance for various reasons. Firstly, it promotes cultural responsiveness within educational settings by acknowledging and respecting the valuable cultural knowledge, perspectives, and experiences that bilingual and Latinx parents bring. This inclusion enables the creation of a curriculum and learning environment that recognizes and appreciates the diverse backgrounds and identities of students. Secondly, involving Latinx parents fosters stronger partnerships among parents, teachers, and schools. When Latinx parents feel welcomed and included in their children's education, they are more likely to actively engage in supporting their child's academic progress, attending school events, and collaborating with teachers.

Furthermore, Latinx parents who are proficient in Spanish, or both English and Spanish, can significantly contribute to bilingual education efforts, particularly in the realm of mathematics. They play a vital role in supporting their children's language and literacy development in both languages, reinforcing language skills, promoting cultural understanding, supporting mathematical learning, and fostering biliteracy. Latinx parents also provide essential academic support to their children, assisting with homework, providing study guidance, and advocating for their educational needs. Their involvement in their children's academic journey can contribute to improved academic outcomes and cultivate a stronger sense of self-efficacy in their children, especially in mathematical proficiency. Additionally, incorporating Latinx parents in education empowers them as active participants and decision-makers in their children's schooling. It allows them to have a voice in shaping educational policies, practices, and programs that directly impact their communities. Moreover, the visibility and engagement of Latinx parents in education serve as positive role models and representation for Latinx students, reinforcing their sense of belonging and pride in their heritage.

In conclusion, the incorporation of Latinx parents in education holds the potential to foster a more inclusive, equitable, and enriching educational experience for students and the wider school community. This study sheds light on potential avenues for collaboration between parents and educators within educational institutions, emphasizing the importance of creating an environment that recognizes and values the cultural wealth contributed by parents (Yosso, 2005).

REFERENCES

Cannon, J., & Ginsburg, H. (2008). "Doing the math": Maternal beliefs about early mathematics versus language learning. *Early Education & Development, 19*(2), 238–260. https://doi.org/10.1080/10409280801963913

Chang, M., Choi, N., & Kim, S. (2015). School involvement of parents of linguistic and racial minorities and their children's mathematics performance. *Educational Research and Evaluation, 21*(3), 209–231. https://doi.org/10.1080/13803611.2015.1034283

Civil, M. (2016). STEM learning research through a funds of knowledge lens. *Cultural Studies of Science Education, 11*(1), 41–59. http://doi.org/10.1007/s11422-014-9648-2

Civil, M., & Quintos, B. (2022). Mothers and children doing mathematics together: Implications for teacher learning. *The Teachers College Record, 124*(5), 13–29. https://doi.org/10.1177/01614681221105008

Clarke, V., & Braun, V. (2013). *Successful qualitative research: A practical guide for beginners.* SAGE.

Colegrove, K. S.-S., & Krause, G. H. (2017). "Lo hacen tan complicado": Bridging the perspectives and expectations of mathematics instruction of Latino immigrant parents. *Bilingual Research Journal, 40*(2), 187–204. https://doi.org/10.1080/15235882.2017.1310679

Dearing, E., Kreider, H., Simpkins, S., & Weiss, H. B. (2006). Family involvement in school and low-income children's literacy: Longitudinal associations between and within families. *Journal of Educational Psychology, 98*(4), 653–664. https://doi.org/10.1037/0022-0663.98.4.653

Deflorio, L., & Beliakoff, A. (2015). Socioeconomic status and preschoolers' mathematical knowledge: The contribution of home activities and parent beliefs. *Early Education and Development, 26*(3), 319–341. https://doi.org/10.1080/10409289.2015.968239

Duncan, G. J., Dowsett, C. J., Claessens, A., Magnuson, K., Huston, A. C., Klebanov, P., Japel, C. (2007). School readiness and later achievement. *Developmental Psychology, 43*(6), 1428–1446. https://doi.org/10.1037/0012-1649.43.6.1428

Englund, M. M., Luckner, A. E., Whaley, G. J. L., & Egeland, B. (2004). Children's achievement in early elementary school: Longitudinal effects of parental involvement, expectations, and quality of assistance. *Journal of Educational Psychology, 96*(4), 723–730. http://dx.doi.org/10.1037/0022-0663.96.4.723

Hart, B., & Risley, T. R. (1995). *Meaningful differences in the everyday experience of young American children.* Paul H Brookes Publishing.

Kleemans, T., Peeters, M., Segers, E., & Verhoeven, L. (2012). Child and home predictors of early numeracy skills in kindergarten. *Early Childhood Research Quarterly, 27*(3), 471–477. http://dx.doi.org/10.1016/j.ecresq.2011.12.004

Lave, J., & Wenger, E. (1991). *Situated learning: Legitimate peripheral participation.* Cambridge University Press.

Moll, L. C., Amanti, C., Neff, D., & Gonzalez, N. (1992). Funds of knowledge for teaching: Using a qualitative approach to connect homes and classrooms. *Theory Into Practice, 31*(2,), 132–141.

Nievar, M. A., Jacobson, A., Chen, Q., Johnson, U., & Dier, S. (2011). Impact of HIPPY on home learning environments of Latino families. *Early Childhood Research Quarterly, 26*(3), 268–277. http://dx.doi.org/10.1016/j.ecresq.2011.01.002

Segers, E., Kleemans, T., & Verhoeven, L. (2015). Role of parent literacy and numeracy expectations and activities in predicting early numeracy skills. *Mathematical Thinking and Learning,* 17, 219–236. https://doi.org/10.1080/10986065.2015.1016819

Stoehr, K., & Patel, P. (2018). Meaningful mathematical discussions that matter. In S. Crespo, S. Caledón-Pattichis, & M. Civil (Eds.), *Access and equity, promoting high-quality mathematics* (Chapter 7). National Council of Teachers of Mathematics.

Stoehr, K., Salazar, F., & Civil, M. (2022). The power of mothers and teachers engaging in a mathematics bilingual collaboration. *Teachers College Record, 124*(5), 30–48. https://doi.org/10.1177/01614681221103947

Watts, T. W., Duncan, G. J., Siegler, R. S., & Davis-Kean, P. E. (2014). What's past is prologue: Relations between early mathematics knowledge and high school achievement. *Educational Researcher, 43*(7), 352–360. https://doi.org/10.3102/0013189X14553660

Wen, X., Bulotsky-Shearer, R. J., Hahs-Vaughn, D. L., & Korfmacher, J. (2012). Head Start program quality: Examination of classroom quality and parent involvement in predicting children's vocabulary, literacy, and mathematics achievement trajectories. *Early Childhood Research Quarterly, 27*(4), 640–653. https://doi.org/10.1016/j.ecresq.2012.01.004

Wilder, S. (2017). Parental involvement in mathematics: Giving parents a voice. *Education 45*(1), 104–121. https://doi.org/10.1080/03004279.2015.1058407

Yosso, T. (2005). Whose culture has capital? A critical race theory discussion of community cultural wealth. *Race, Ethnicity and Education, 8,* 69–91. https://doi.org/10.1080/1361332052000341006

INDEX

Mathematics Instruction in Dual Language Classrooms, pages 191–197

C

D

N

O

P

R

S